T0321369

Preventing Medical Device Recalls

Preventing Medical Device Recalls

DEV RAHEJA

CRC Press
Taylor & Francis Group
Boca Raton London New York

CRC Press is an imprint of the
Taylor & Francis Group, an **informa** business

A PRODUCTIVITY PRESS BOOK

CRC Press
Taylor & Francis Group
6000 Broken Sound Parkway NW, Suite 300
Boca Raton, FL 33487-2742

© 2015 by Taylor & Francis Group, LLC
CRC Press is an imprint of Taylor & Francis Group, an Informa business

No claim to original U.S. Government works

Printed on acid-free paper
Version Date: 20140611

International Standard Book Number-13: 978-1-4665-6822-8 (Hardback)

Library of Congress Cataloging-in-Publication Data

Raheja, Dev, author.
  Preventing medical device recalls / Dev Raheja.
    p. ; cm.
  Includes bibliographical references and index.
  ISBN 978-1-4665-6822-8 (hardback : alk. paper)
  I. Title.
  [DNLM: 1. Medical Device Recalls--United States. 2. Equipment Safety--United States. 3. Product Surveillance, Postmarketing--United States. 4. Risk Assessment--United States. WA 289]

R859.7.S43
610.28'9--dc23   2014016847

Visit the Taylor & Francis Web site at
http://www.taylorandfrancis.com

and the CRC Press Web site at
http://www.crcpress.com

This book is dedicated to the managers and engineers who do their best so nothing bad can happen to patients.

Contents

Preface

A critical and sometimes overlooked aspect of preventing medical device recalls is the inability to implement systems thinking. Systems thinking is not a panacea that can prevent every mistake, but it is the best-known tool that helps us predict hidden risks and see many robust solutions to eliminate risks. Unlike the limited single-focused cause-and-effect approach, systems thinking is about the relationship of individual causes to system causes, interdependencies of the knowledge of cross-functional teams, interdependencies of the knowledge of manufacturers and device users, and understanding what you don't know.

No individual has had more influence on quality management than Dr. W. Edwards Deming. His Theory of Profound Knowledge was the chief reason for the success of Japanese businesses such as Toyota, Fuji, and Sony, and was the basis of systems thinking. He taught them that a system is lot more than the sum of its parts. Everyone doing their job right without understanding the interactions and interdependencies among subsystems and components is not enough. He preached that management is accountable for 85% of problems—*not* workers—because of a lack of systems thinking. His goal of quality was to exceed customer needs through system improvement. This book provides a structure for systems thinking to most efficiently prevent device recalls.

Acknowledgment

I am grateful to Dr. Peter Pronovost, senior vice president of Patient Safety and medical director of the Center of Innovation in Quality Patient Care at Johns Hopkins Hospital, for giving me the inspiration to write this book. The idea started when he and I wrote the article "Protecting Patients from Dangers in Medical Devices." That article is reprinted in Appendix A of this book. Special thanks also to physicians and healthcare professionals Dr. David Brown and Dr. Maria Escano (surgeon and coauthor of several articles on patient safety). Thanks also to Richard Ramirez, Avinash Konkani, Joseph Schnur, and Gauri Raheja for making valuable suggestions.

I am especially thankful to Oriel STAT A MATRIX for opportunities to teach medical device courses in their public seminars for many years. Thanks also to my former employer, GE Healthcare, for initiating my interest in public safety; to Siemens Medical System for using my consulting services for two years in Stockholm, Sweden; and to several universities (University of California–Los Angeles, George Washington University, University of Wisconsin, University of Maryland, and University of Alabama) for opportunities to teach short courses on safety and reliability. Thanks to my family, especially Hema, for excellent support.

About the Author

Dev Raheja, MS, CSP, has been an international risk management and quality assurance consultant in the healthcare, medical device, and aerospace industries for more than twenty-five years. He applies evidence-based safety techniques from a variety of industries to healthcare.

He is a trainer, and author of the books *Safer Hospital Care, Assurance Technologies: Principles and Practices,* and *Design for Reliability.* He shows clients how to create elegant solutions using creativity and innovation. Being a true international consultant, he has conducted training in several countries including Sweden, Australia, Japan, Germany, the United Kingdom,

Singapore, Taiwan, South Africa, Finland, and Brazil. He helped a major company in the midwestern United States avoid going out of business and become a world leader by eliminating safety mishaps.

Prior to becoming a consultant in 1982, he worked at GE Healthcare as supervisor of quality assurance and manager of manufacturing, and at Booz-Allen & Hamilton as a risk management consultant for the nuclear and mass transportation industry.

Raheja served as adjunct professor at the University of Maryland for five years in its PhD program in reliability engineering, and is currently an adjunct professor at Florida Tech for its BBA degree in healthcare management. He is an associate editor for healthcare safety for the *Journal of System Safety*, and teaches webinars on medical device safety and reliability. He has received several industry awards including the Scientific Achievement Award and Educator-of-the Year Award from the System Safety Society and the Austin Bonis Reliability Education Award from the American Society for Quality.

He served as part of the first group of examiners for the Malcolm Baldrige National Quality Award, and served for fifteen years on the board of directors of the Annual Reliability and Maintainability Conference sponsored by ten engineering societies. Currently he is the member of the Institute of Electrical and Electronics Engineers (IEEE), Association for the Advancement of Medical Instrumentation (AAMI), the American Society of Patient Safety Professionals, and the American College of Healthcare Executives.

Raheja majored in human factors engineering as a part of his master's degree in industrial engineering, is a Certified Safety Professional through the Board of Certified Safety Professionals, and serves as the chairman of the Design for Reliability Committee of the IEEE.

Chapter 1

Introduction to Medical Device Requirements

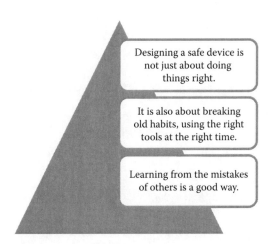

Designing a safe device is not just about doing things right.

It is also about breaking old habits, using the right tools at the right time.

Learning from the mistakes of others is a good way.

Introduction

Medical device recalls can cost billions. The number of medical device products affected by recalls hit an eight-quarter high in the second quarter of 2012, exceeding 100 million units. Perhaps most striking is the fact that of the 140 medical device companies affected by recalls in the 90-day period in that quarter, one third faced multiple recalls. More than

10 million units were affected. This is a reminder that recalls can happen to any company—especially as more and more companies turn to outsourcing [1]. A division of a highly reputable company paid over $4 billion just to settle 8,000 lawsuits related to a recalled hip joint device that caused severe pain and injury to patients due to generation of metallic particles. In addition, the company paid the surgical expenses of those affected.

While most standards are helpful, they can be considered only minimum requirements where self-interest prevails [2]. Even in healthcare, where compliance with standards is at its highest level, more patients die from medical mistakes each week than would be involved in a jumbo jet crash. Understanding how mishaps can happen before they happen and preventing them by design are key. This is the intent of this volume.

The direct financial costs of such mishaps may include

- Cost of product removal, investigation, and corrective medical services to recover from harm.
- Administration of the recall (including legal fees, communication with patients and hospitals, contractor fees, and higher insurance costs).
- Product replacement and priority shipping costs.
- Shutdown of the facility for the duration determined by the U.S. Food and Drug Administration (FDA). (A company was not allowed to sell its radiology system for two years; a handy-wipe company was shut down for three months.)

The indirect costs may include

- Competitive disadvantage in the marketplace.
- Bad reputation.
- Warranty claims and other customer issues (including product liability claims).
- Workplace stress.

- Shutdown of facilities.
- Criminal charges against executives. (Four executives of a company were ordered to spend six months in jail for authorizing an unapproved use.)

The Challenges

The challenges faced by medical device manufacturers in bringing safe, reliable products to market in a timely fashion that are also low in cost over the product's life cycle are increasing rapidly. The increasing complexity and software criticality of medical devices are driving the need for a proactive practice of risk prevention. However, theoretical knowledge is not enough. This book covers paradigms for proactive thinking and doing. It presents examples from the author's experience of over thirty years in the medical device, consumer, and aerospace industries.

The number of FDA product recalls in the United States increased from 527 in 2003 to 845 in 2008. The FDA received more than 10,000 complaints per year on infusion pumps alone during this period [3]. Some pump manufacturers say that most problems occur when a nurse or healthcare worker enters the wrong data accidentally. However, FDA officials found that many deaths and injuries related to the devices were caused by product design and engineering flaws, rather than user error. This book shows how to compensate for user errors in the design. "First do no harm" should be the priority of every design. The number of recalls per year at the time of this writing is at an all-time high. The book is written for U.S. markets, but the principles described are universal.

Sources of Errors

Incomplete and vague specifications are a major reason for product recalls. Figure 1.1 shows the distribution of the causes

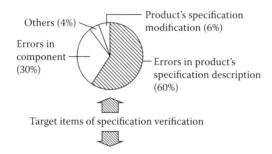

Figure 1.1   Distribution of the causes of product failures and malfunctions.

of product failures [4] in most industries. The author's more than thirty years of experience with medical devices supports this.

This figure shows that about 60% of problems are in the product specification, and 30% of the errors are in components, which are usually errors in component specifications, such as no reference to reliability, durability, or extreme environments in which the component has to work, and safety-critical features in the component.

Beyond specification, the Institute of Medicine (IOM), in their landmark report "To Err Is Human," claims that most errors are caused by faulty systems, processes, and conditions. These are often called management errors.. This is true for manufacturing companies also. Dr. Edward Deming, the father of Japanese quality fame, says that 85% of problems are management problems—mostly inadequate procedures and inability to implement the procedures correctly. Companies usually get sufficient warnings about failure to implement good basic design and production procedures during FDA audits. Such companies are vulnerable to product recall. The secret to preventing medical device recalls is to follow good evidence-based practices related to specifications control, design control, design for process reliability, production validation, and design validation using accelerated life tests. Here are some common FDA warnings given to some reputable companies [5].

- Failure to implement complaint handling procedures
- Failure to establish and maintain procedures for implementing corrective and preventive actions
- Failure to establish and maintain procedures for acceptance of incoming product
- Failure to adequately validate with a high degree of assurance and approve according to established procedures, a process that cannot be fully verified by subsequent inspection and testing
- Failure to establish and maintain adequate procedures for validating device design
- Failure to maintain procedures to ensure that all product purchased or otherwise received conforms to specifications
- Failure to establish adequate procedures for quality audits
- Failure to establish and conduct adequate procedures to control the design of the device in order to ensure that specified design requirements are met
- Failure to establish sampling plans based on a valid statistical rationale
- Failure to establish and maintain adequate procedures for finished device acceptance to ensure that each production run, lot, or batch meets acceptance criteria
- Failure to ensure that only those devices that are approved for release are distributed
- Failure to process complaints in a uniform and timely manner

These warnings show that the symptoms of potential recalls are evident, but they are handled reactively instead of looking proactively at the big picture. This book goes a step further in preventing recalls. It shows how to identify and prevent possible failures and malfunctions while the design is still on paper, resulting in prevention of costly engineering changes and harm to patients. It also helps in preparing for legal product liability prevention. It complements regulatory requirements, which are often not prescriptive. They are a minimum standard of performance, and do not represent comprehensive requirements to prevent recalls. They serve as a good checklist for essential protocols for device efficacy.

Dr. W. Edward Deming used to say, "It is not enough to just *do* your best or work *hard*. You must know what to *work* on." In this author's view, the following three things are the right things to do and are the focus of this book:

1. Identify as many potential harmful scenarios as possible and mitigate them prior to releasing the specification (see Chapters 2, 3, and 4).
2. Identify life-cycle defects and malfunctions, make design changes to prevent them, and validate life-cycle performance (see Chapters 5, 6, 7, 8, 9, 10, and 11).
3. Identify management gaps and fix them (see Chapters 12, 13, 14, and 15).

It is extremely important that a device manufacturer's product development team understand its customers' requirements before starting the product design; if a product does not meet users' expectations of safety, even during the occasional foreseeable misuse, the device can be subject to recall. Note that a hospital or a patient is not the only customer. For a medical device company, contract manufacturers and suppliers (who need good specifications on their components and identification of features critical to quality and safety), consumers who buy the equipment for personal use (e.g., home healthcare providers, diabetic patients, etc.), and technicians who service the equipment (they need to understand the design and its critical performance requirements) are customers, too. Hence, the product development team should ensure that its members understand the safety needs of all the different types of stakeholders. In this chapter we will cover important regulatory requirements, especially for those entering the medical device field.

Understanding the Science of Safety

The first step, before we start developing a product, is to understand the science of safety. Without this understanding, the chance of a recall is high. The science of safety is about protecting patients from harm by understanding human errors and equipment malfunctions. Protection is achieved by preventing harm from happening by designing very reliable devices capable of detecting anomalies before harm is done. The methods for designing in this way are covered throughout the book.

Overview of FDA Quality System Regulation

FDA Quality System Regulation 21 CFR 820 has been in effect since 1997 (see Appendix B or www.fda.gov). It requires manufacturers to have a quality management system (QMS) to address risks in the devices, the complexity in the device, the complexity in manufacturing processes, and the complexity of the entire organization. Each manufacturer, including manufacturers of *in vitro* diagnostic devices, is required to establish procedures for managing these issues and treating them as living documents; that is, updating them as new risks develop and documenting the data to ensure the effectiveness of their implementation. It applies to any organization (or person) that designs, manufactures, fabricates, assembles, or processes a finished device. Manufacturers include, but are not limited to, those who perform the functions of contracted sterilization, installation, relabeling, remanufacturing, repacking, or specification development, and initial distributors of foreign entities performing these functions [6]. This reference also highlights the subsystems that must be managed, as shown in the Figure 1.2. Management itself is a subsystem of the entire system of getting high-quality, safe products to users. Its responsibility is to develop quality policies and make sure that employees understand and implement them well. Other responsibilities include providing adequate training

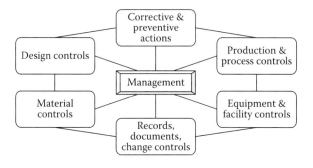

Figure 1.2   Seven subsystems that must be integrated to achieve the goal of a safe and effective product. (From Kimberly A. Trautman, "Presentation: Quality System Regulation 21 CFR 820 Basic Introduction," U.S. Food and Drug Administration, March 17, 2011, http://www.fda.gov/MedicalDevices/ResourcesforYou/Industry/ ucm126252.htm)

and resources, assuring audits, participating in reviews, and appointing a management representative to facilitate the efficiency and effectiveness of the system.

The FDA requires a good design control process in which the needed inputs to design are identified prior to the start of a design and the expected outputs are verified. A design can be divided into several phases depending on complexity. Then design output for a phase becomes the design input for the next phase. Activities such as design reviews, changes, verification, and validation are a part of design control. Verification applies to verifying that the inputs to the design are adequate and so are the outputs. Validation usually refers to final testing to ensure the product meets the implied user needs. All the findings and actions are required to be documented in the Design History File.

A manufacturer must also demonstrate that controls in manufacturing, purchasing, shipping, handling, storage, distribution, installation, maintenance, and disposal are in place. Safety analysis for all these activities must be done so that no latent hazards are introduced into the device. Principles of verification and validation also apply to the design of a manufacturing process.

Use of statistical tools is required to document that the established controls are effective over time. Each new defect, complaint, or device failure is supposed to be processed through a formal corrective and preventive action (CAPA) system and should be traceable to its original finding. The entire FDA Quality System Regulation is available in Appendix B of this volume.

Overview of Risk Management Standard ISO 14971

International Organization for Standardization (ISO) 14971 is an international voluntary risk management standard [7]. Since this is more prescriptive, it complements worldwide regulations. All Class III devices (highest risk) should make use of this standard in its entirety. Class II and Class I device makers can choose what is appropriate for their devices. They can also use this standard to prevent defects and failures as a good business practice.

The standard describes risk as an aggregate of the probability of harm and its severity. It is a good tutorial on risk analysis, risk evaluation, and risk control. It makes us aware of not only identifying hazards, but also hazardous situations. Both are present in every harm scenario. To prevent risk, one must either design out the hazard or prevent the hazardous situation. The standard covers what to do if there are residual risks that cannot be mitigated, which include assessing whether the product should be marketed in the interest of public health.

It covers how to develop a risk management plan, what tools to use for risk analysis, and precedence for risk control strategies. The risk analysis techniques recommended are as follows [7]:

■ Preliminary Hazard Analysis (PHA)
■ Fault Tree Analysis (FTA)
■ Failure Mode and Effects Analysis (FMEA)

- Hazard and Operability Study (HAZOP)
- Hazard Analysis and Critical Control Points (HACCP)

This standard gives a brief description of the techniques, but one needs in-depth understanding. This book goes deeper into these techniques and provides hands-on examples.

One of the very useful features is that it gives guidance in detail on risk management for *in vitro* diagnostic medical devices in Annex H of the standard. These devices are intended for use in the collection, preparation, and pathological examination of samples taken from the human body. The devices include reagents, instruments, software, sample collection devices and receptacles, calibrators, control materials, and related accessories. The guidance includes examples of possible use errors by laboratory personnel, healthcare providers, and by patients doing self-testing. Also included are examples of known and foreseeable hazards.

Overview of FDA Device Approval Process

There are three classes of devices. Classes are based on the risk the device poses to the patient and the user's degree of control necessary to assure that the various types of devices are safe and effective [8]. The classification depends on the *intended use* of the device and upon *indications for use*. Class I devices pose the lowest risk and Class III devices possess the greatest risk. Class III devices support or sustain human life, or are of substantial importance in preventing impairment of human health, or present a potential, unreasonable risk of illness or injury. All Class III devices are subject to 510(k) premarket approval requirements, a process to ensure the safety and effectiveness of the device [8]. Class I devices are not intended to help support or sustain life or be substantially important in preventing impairment to human health, and may not present an unreasonable risk of illness or injury.

Most Class I devices are exempt from premarket notification and a few are also exempted from most good manufacturing practices regulations. Examples of Class I devices include elastic bandages, examination gloves, and hand-held surgical instruments. If a device is classified as Class I or II and if it is not exempt, a 510(k) will be required for marketing.

Devices in Class II are held to a higher level of assurance than Class I devices and are designed to perform as indicated without causing injury or harm to patient or user. Examples of Class II devices include powered wheelchairs, infusion pumps, and surgical drapes.

All devices are required to implement general controls before marketing, such as application of good manufacturing practices, maintenance of required records and reports, prevention of product adulteration, and prevention of misbranding. All devices classified as exempt are subject to limitations on exemptions. The guidelines for all these requirements are available at the FDA website, http://www.fda.gov

Overview of Regulatory Requirements for Clinical Trials

The FDA is responsible for protecting the public health by assuring safety, efficacy, and security of human and veterinary drugs, biological products, medical devices, the food supply, cosmetics, and products that emit radiation. The FDA is also responsible for advancing the public health by helping speed innovation and getting the public the accurate, science-based information they need. Clinical trials are required for Class III devices to determine if a new device's benefits to the public outweigh the risks.

In determining the safety and effectiveness of a device, the FDA considers the following, among other relevant factors (the following text is quoted directly from, the FDA *PMA Clinical Studies* [9]:

1. The persons for whose use the device is represented or intended
2. The conditions of use for the device, including conditions of use prescribed, recommended, or suggested in the labeling or advertising of the device, and other intended conditions of use
3. The probable benefit to health from the use of the device weighed against any probable injury or illness from such use
4. The reliability of the device

Although the manufacturer may submit any form of evidence to the FDA in an attempt to substantiate the safety, reliability, and effectiveness of a device, the FDA relies upon valid scientific evidence to determine whether there is reasonable assurance that the device is safe and effective. Valid scientific evidence is evidence from well-controlled investigations, partially controlled studies, objective trials and studies without matched controls, well-documented case histories conducted by qualified experts, and reports of significant human experience with a marketed device, from which it can fairly and responsibly be concluded by qualified experts that there is reasonable assurance of the safety and effectiveness of a device under its conditions of use. The evidence required may vary according to the characteristics of the device, its conditions of use, the existence and adequacy of warnings and other restrictions, and the extent of experience with its use. Isolated case reports, random experience, reports lacking sufficient details to permit scientific evaluation, and unsubstantiated opinions are not regarded as valid scientific evidence to show safety or effectiveness.

The valid scientific evidence used to determine the safety of a device must adequately demonstrate the absence of unreasonable risk of illness, new risks, or injury associated with the use of the device for its intended uses and conditions of use. The premarket approval application must include a discussion of the conclusions drawn from studies conducted with the medical device. FDA does not prescribe specific statistical analyses for given devices and/or situations. All statistical analyses used in an investigation should be appropriate to the analytical purpose, and thoroughly documented.

The discussion should demonstrate that the data and information in the application constitute valid scientific evidence and provide reasonable assurance that the device is safe and effective for its intended use.

The analysis should include the following:

- Summary of results (graphs are helpful).
- Summary of the study subjects including the number of subjects who have prematurely discontinued participation. (Include patient tree and spreadsheets to provide full accounting of all study subjects including controls and drop-outs, as appropriate.)
- Description of events potentially affecting study success (e.g., difficulties enrolling patients; changes in key personnel; discontinuation of participation by subjects and investigators).
- Summary of anticipated and unanticipated adverse effects.
- Description of any deviations from the investigational plan by investigators.
- Discussion of any missing data and how it impacts the study.
- Description of method of statistical analyses used; describe how any assumptions required in the statistical analysis were validated.
- Comparison of results to success/failure criteria.
- Conclusions drawn from study, relate back to indications for use and how the data supports each indication.

Summary

Learning from the mistakes of others is a good way to prevent medical device recalls. Go to the FDA website (http://www.fda.gov) and look up the warning letters, adverse event reports, and reasons for recall of devices. You will discover that even though the products recalled are not similar to your device, the causes may very well apply to your device. Understand the design, manufacturing, and servicing risks. Mitigate risks while writing specifications and developing the concept design. Such a strategy can save your company millions of dollars by preventing recalls and lawsuits. Besides, there are immediate savings from preventing failures; you can drastically reduce warranty costs and those of engineering changes. In addition, be very familiar with the Quality System Regulation 21 CFR 820 (Appendix B) and the ISO 14971 Standard for Risk Management.

References

1. Hartford, Jamie. "Medical Device Recalls Hit 2-Year High in Q2 2012," Medical Devices and Diagnostics Industry, August 23, 2012, http://www.mddionline.com/article/medical-device-recalls-hit-2-year-high-q2-2012 (accessed March 20, 2014).
2. Peters, George A., and Peters, Barbara J. *Human Safety, Volume 2*. Santa Monica, CA: Peters & Peters, 2013.
3. Grammatech, *FDA Recommends Static Analysis for Medical Devices*, 2010, http://www.grammatech.com/images/pdf/grammatech-fda-case-study.pdf (accessed May 21, 2013).
4. Raheja, Dev, and Allocco, Michael. *Assurance Technologies Principles and Practices*. New York: Wiley, 2006.
5. U.S. Food and Drug Administration (FDA), FDA's Electronic Reading Room: Warning Letters, http://www.accessdata.fda.gov/scripts/warningletters/wlFilterByCompany.cfm (accessed April 19, 2013).
6. Kimberly A. Trautman, "Presentation: Quality System Regulation 21 CFR 820 Basic Introduction," U.S. Food and Drug Administration, March 17, 2011, http://www.fda.gov/MedicalDevices/ResourcesforYou/Industry/ucm126252.htm
7. ANSI/AAMI/ISO 14971:2007, *Medical Devices-Application of Risk Management to Medical Devices*, Washington, DC: American National Standards Institute.

8. U.S. Food and Drug Administration (FDA), *PMA Approvals*, http://www.fda.gov/MedicalDevices/ DeviceRegulationandGuidance/Overview/ClassifyYourDevice/ (accessed April 19, 2013).

9. U.S. Food and Drug Administration (FDA), *PMA Clinical Studies*, http://www.fda.gov/MedicalDevices/DeviceRegulationandGuidance/ HowtoMarketYourDevice/PremarketSubmissions/ PremarketApprovalPMA/ucm050419.htm (accessed 2013).

Chapter 2

Preventing Recalls during Specification Writing

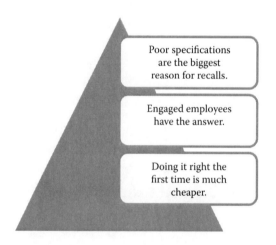

Poor specifications are the biggest reason for recalls.

Engaged employees have the answer.

Doing it right the first time is much cheaper.

Introduction

Once upon a time, the designer was in charge of everything that could go wrong in a product's performance. Today, products, manufacturing processes, supply chain controls, and product users have become disengaged in many companies; therefore, most design engineers are far removed from many things going wrong. Take the example of a new x-ray system designed to be very user-friendly. It required the user to press "1" on the keyboard for regular chest radiation, and press "2" for very high radiation for therapeutic applications. It was certainly user-friendly because it did not require any training, but it turned out to be patient-unfriendly. Ten patients died in ten hospitals in the first week of the product introduction. The radiology technicians made the predictable error of inadvertently pressing the "2" key instead of the "1" key. We do this all the time; we press the key next to the one we should have pressed. The U.S. Food and Drug Administration (FDA) required the company to stop selling the system for two years until there was proof that a new design was safe. Every technical person and senior management took training on designing for safety. The question is, why did the design engineer overlook such fundamental safety issues? The answer is that no one person can capture all the requirements. Even a team of intelligent cross-functional members cannot capture them all, but the company can capture most of the requirements with the structured methods covered in this chapter.

Conduct Requirements Analysis to Identify Missing Requirements

The most efficient way to prevent recalls is to identify as many potential problems related to safety and efficacy as early as possible using the risk management tools covered in International Organization for Standardization (ISO) 14971 described in Chapter 1. This needs to be done before approving specifications because there will be major changes after using the analysis tools. Use any other tool if you can, such as brainstorming; ask what-if questions, and search competitor malfunctions on the Internet. Mitigate the high risks and include the solutions as performance functions in specifications.

Good requirements management is critical to success for all products. To ensure that the product meets customer needs, expressed and implied, we need to define measurable requirements in the specifications. The challenge starts when the team agrees to the specifications without challenging them. That is the main reason for many missing requirements. Therefore, a sound structure for the design process and team management must be in place. At a minimum, the specification should challenge the user's needs, the environment of use including any interacting devices, and inadvertent misuse by patients or caregivers. This is conceptually shown in Figure 2.1.

The design process should also include a system to recognize employees who are fully engaged and creative. According to the Gallup organization, out of a wide range of industries, less than 30% of employees are engaged and willing to walk the extra mile for the good of the company [1]. Management can reward teams for failure-free performance, which will engage more employees. Engaged employees know the problems and they try to solve them, but disengaged or unengaged peers get in their way.

The sources of most failures are incomplete, ambiguous, and poorly defined device specifications. They result in expensive engineering changes later, one at a time; this is called *scope creep*. Often, robust changes cannot be made because the project is delayed and there may not be sufficient budget for new features.

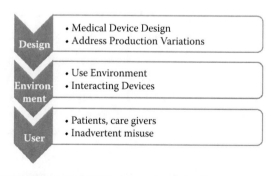

Figure 2.1   Considerations in the specifications.

Look particularly for missing functions in the specifications. Usually, there is hardly anything about modularity to minimize interactions, increase reliability, safety, serviceability, improve logistics, and take into account human factors, testability, diagnostics capability, and prevention of old failures in current design. Very few specifications address even obvious requirements, such as the requirements for internal interfaces, external interfaces, user–hardware interfaces, and user–software interfaces; and do not address how the product should behave if and when there is an unexpected event, such as the device shutting down from a power failure.

Developing a good specification is an iterative process with inputs from the customer and the entities that are downstream in the process. Those who are trying to design around a faulty specification should expect only a faulty design. Unfortunately, most companies discover problems when the design is already in production. At that stage, there are usually no resources and no time for major design changes. The only thing a company can do is to hope for a band-aid solution and commit to do it better the next time.

To identify missing functions in a specification, a tool called *requirements analysis* is necessary. This analysis is always conducted by a cross-functional team. At least one member from each discipline should be present, such as R&D, design, quality, reliability, safety, manufacturing, field service, marketing, and if possible, a customer representative. As shown in Chapter 1, as much as 60% of necessary requirements are missing in new device specifications. Therefore, if the specifications contain only 40% of the necessary features, we open ourselves to a good chance of product recall. Therefore, writing accurate and comprehensive performance specifications is a prerequisite for a safe and reliable design. The author's interviews with those attending his safety courses reveal that the troubleshooting technicians on complex electronic products are unable to diagnose about 65% of the problems (cannot duplicate the fault). Obviously, fault isolation requirements in the specifications are necessary for such devices.

To ensure the accuracy and completeness of specifications, only those who have knowledge of what goes into good specifications should approve them. They must ensure that the specifications are clear on what the device should *never* do, however stupid that may sound. For example: There shall be no false positive and negative alarms from the device used in intensive care units in hospitals. (A study in the National Children's Medical Center showed over 85% of alarms were false positives. Nurses were required to check on patients who did not need immediate help instead of attending to patients who were in danger of dying.) The "shall not" specification is not limited to failures. That would be too simple. Try to antici-pate adverse events. They also should never happen.

To find design flaws early, the team must view the system from different angles. You would not buy a house by just looking at the front view. You want to see it from all sides. Similarly, a device concept has to be viewed from at least the following perspectives:

- Functions of the product.
- Undesired functions that should never occur.
- Range of applications.
- Range of environments.
- Active safety (safety process during use of the device).
- Duty cycles during life.
- Reliability.
- Robustness for user/servicing mistakes.
- Logistics requirements to avoid adverse events.
- Manufacturability requirements.
- Internal interface requirements.
- External interface requirements.
- Installation requirements to ensure safe functioning. (An MRI should be accurate once it is installed.)
- Shipping/handling capabilities to keep the device safe.
- Serviceability/diagnostics capabilities.

- Prognostic health monitoring to warn users in case of an anomaly.
- Interoperability with other products.
- Sustainability.
- Potential accidents and abuses.
- Human factors.

In this list, there is a need to explain what a sustainable design does. It is about meeting current needs without compromising the needs of future generations, such as not introducing harm to the environment through pollution or disposing of medical waste in rivers. Many electronic products are not designed for sustainability. They should be designed for reuse or recycling. Unfortunately, not everyone makes an effort to recycle. Virtually all electronics contain toxic materials that can be harmful to people. A lot of this hazardous stuff lives in the circuit boards. In 2008, we generated 3.16 million tons of electronics waste in the United States. According to the EPA, only 13.6% was recycled [2].

Most designers are likely to miss many of these requirements. This is not a new situation, but it can be addressed by inviting experts in these areas to brainstorm. There is no mechanism for customers to specify all of these requirements. Companies that want to be productive will teach customers and solicit their participation as team members. The point is that if we do everything right the first time, we don't have to fix many mistakes later.

Specifications for Safety, Durability, and Reliability

Safety, durability, and reliability are the most important requirements to prevent recalls. Of these, safety requirements are the most difficult to identify. *Safety* can be defined as the condition where no harm will be done to the user. It is very critical that designers understand safety theory, which states that harm cannot occur until at least two things go wrong: First, there must be a source of danger in the use environment and second, there must be an unsafe event, such as a human error, that can trigger the source of danger into harm. The risk management standard ISO 14971 calls them a "hazard" and a "hazardous situation," respectively. Therefore, to identify safety requirements, we must identify these combinations that can result in harm. We must identify all the hazards (sources of harm) in the system consisting of the device hardware, software in the device, use procedures, user behavior, maintenance, and repair behaviors. Then, for each hazard, we need to identify the hazardous situations that transform the hazard into harm. For each combination of the hazard and the hazardous situation, we need to mitigate the risk of either the hazard or the hazardous situation. We don't have to mitigate both, because harm cannot take place if only one of these is present. The mitigation chosen becomes a safety requirement. If there are 70 hazards, we should have 70 new safety requirements!

As an example of developing a safety requirement, the patient-controlled analgesia (PCA) pump shown in Figure 2.2 is designed so that a patient can administer doses of narcotics to control severe pain after surgery. Sometimes, caregivers can inadvertently program the pump incorrectly. Sometimes family members manage the pump for the patient. It is a method of allowing persons in pain to administer their own pain relief. An infusion pump is programmable by doctors and nurses. If it is programmed and functioning as intended, the machine is unlikely to deliver an overdose of medication. Regardless of who is controlling the administration, a *hazard* in this case is the ability to overdose, which can result in severe harm.

The ability to deliver a high dose does not itself result in harm unless a *hazardous situation* occurs, when a patient

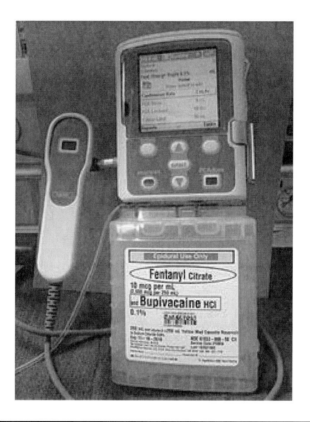

Figure 2.2   An example of a PCA pump.

or family member actually selects a high dose that is beyond the safe limit. The safety requirement is to either prevent the hazard (taking away the patient's ability to increase the dose) or prevent the delivery of a high dose no matter what the patient or the family member does by limiting the maximum amount of medication that can be administered in the equipment design. Since we want the patient to control pain, the wise choice is to limit the maximum medication by design. The specification should include the maximum amount of pain medication that can be delivered by the pump.

Durability is just as important for safety. Instruments and computers are known to fail during critical care and surgeries, putting the life of the patient at risk. A patient who tries to deliver pain medication can experience sudden failure of the device. Durability is defined as the time to the first failure from wear and tear of the equipment components. It is also called the *minimum life*. To have enough durability, the product specification and the component specifications should include the minimum life. The product specification should also include the requirement to inform users to replace the equipment before the minimum life expires (under the heading "logistics requirements"). A preventive maintenance (PM) program for replacing critical parts before they fail would significantly improve the reliability and durability of the device. If the equipment is used in very critical care, an alarm must be included to signal that the equipment (or component) must be replaced when the minimum life is about to expire. The legal system expects that some users will use the equipment beyond the expiry date out of lack of knowledge or because they have no choice. The burden is on the manufacturer to protect patients from adverse events, either through a warning on the device or through an alarm.

Similar to durability, reliability should also be specified for the planned minimum life. The difference between durability and reliability is that reliability failures occur any time while durability-related failures are related to components wearing out. Reliability failures can result from manufacturing mistakes, inadequate processing, inadequate use procedures, or inadequate use.

Make sure to specify high reliability for the device. *Reliability* is defined as the percentage of products that will work for a specified length of time. It is usually quantified; for example, Reliability R for three years = 0.995 means that 99.5% of devices will work for three years. The time in the reliability equation can be specified in hours, months, or years. Don't use mean time between failures (MTBF) to measure reliability. MTBF only applies when the failure rate is constant, which is rare in the first several years of use. Ideally, designers can eliminate the cause of failure and get 100% reliability for critical failures. For example, if a joint between two components is the cause of failure, use a single-piece design. No joint, no failures!

Specification for User Interface and Usability

Many devices in hospitals work together. The output of your device may be an input for a connected device and vice versa. The software interfaces between the two may not be compatible and secure. Designers must brainstorm to identify compatibility and security issues (a change in the connecting device can change the functions of your device) and write specifications to avoid communication failures. In addition, the software and hardware within the device can introduce hazards that make other devices malfunction. Such issues must also be addressed in the specifications.

A user interface is also an issue in safety. The device has to be user-friendly. The basic requirement is that the device be designed so that it does not permit users to make a mistake in using it or alerts the user if a mistake is made. A tool called *Usability Hazard Analysis* should be used to identify user hazards in the device. This is a brainstorming tool and solutions are assigned to prevent or control each hazard. These solutions become a part of the device specifications. See Figure 2.3 for such an analysis of an automated ventilator [3].

Specification for Maintainability

Accidents can happen because maintenance is not performed correctly or not performed on time. Inaccurate settings of MRI machines or improper calibrations of other devices are not uncommon. FDA recall Z-0165-2008, *Class 2 Recall ONCOR Expression*, suggests that miscalibration can affect the beam profile of a radiation therapy system [4]. Therefore, the product specifications must provide safeguards against such incidences.

USABILITY HAZARD ANALYSIS

Product: Automated Ventilator

Use Profile

WHAT CAN GO WRONG WITH NORMAL USE? (Include software)

- Battery runs out.
- Pump stops working.
- Patient circuit unhooked.
- Circuit blockage.
- Pressure delivered is too high or too low.
- Pressure sensor failure.
- User interface failure.
- Loss of ground affecting other equipment.

WHAT CAN GO WRONG IN ABNORMAL SITUATIONS?

- Prolonged exposure to high temperature in desert environment.
- Malfunction in prolonged freezing temperatures.
- Unusual power surges.
- Water intrusion resulting in electrical malfunction.

HOW CAN THE PRODUCT BE MISUSED?

- Someone turns off the alarms.
- Someone turns off the LEDs.
- Device stored in extreme environments.

HOW CAN THE PRODUCT BE ABUSED?

- Product carelessly dropped.
- Gets thrown with extra force during transportation and shipping.
- Placement of heavy materials on top of the device during transportation.
- Fine sand intrusion in a desert environment.

Mitigation Strategies: For each of the above risks, choose one or more of the four strategies:

a) Eliminate risk, (b) Fault tolerance, (c) Fail safely, (d) Early warning, and (e) Robust training.

Figure 2.3   An example of Usability Analysis.

Maintainability is defined as the ability to perform maintenance in the specified time, but a better definition would be designing to minimize maintenance. Ideally, the device should not require maintenance at all. This is done by eliminating the failure modes that require maintenance. You will see several examples of completely eliminating failure modes in subsequent chapters. However, if maintenance cannot be eliminated, then we must identify hazards in the maintenance process and try to eliminate them. For example, anesthesia equipment for hospitals contained identical hoses for oxygen supply and nitrous oxide supply. The hazard was that both the hoses had the same fitting, which can allow a distracted maintenance technician to connect a hose to the wrong gas supply (a hazard). What should go into the equipment specification? If you did some brainstorming, the solution is simple: The hose fitting for each hose must be unique so it can only fit the correct gas supply. Many companies have taken this approach after a few patients went into a coma after being given the wrong gas.

Specification for Prognostics

In most products, unexpected behaviors can happen such as the computer slows down, the device malfunctions because of current leakage, someone presses the wrong button on the device, or someone disconnects the device. For example, a housekeeper in a hospital killed two patients separately who were on life-support devices. She wanted to connect the vacuum cleaner to the power source. She unplugged the patient's life support system without realizing the severity of her actions. She plugged in the life-support system after cleaning without knowing that the patient had died because of the removal of the power source. The hospital could not establish the cause of the death until the same incident happened again to another patient. In any such unexpected situations, we must safeguard the patient from harm by designing the device to fail gracefully and default to a safe setting. The system should have been intelligent enough to immediately call the red team or have battery backup for the life-support system. The same incident can happen in the case of a power failure from tornadoes, storms, earthquakes, and so on.

The question is, how does a designer know about such situations? The use of tools such as Preliminary Hazard Analysis (PHA) with the hospital staff can help discover many such anomalies. This tool is covered later in Chapter 3. Sudden loss of power can cause hazards in electronic health records, infusion pumps, and ventilators. Such devices should contain a fail-safe default.

Specification for Safe Software

The most frequently recalled devices were linear accelerators [5]. The FDA data demonstrates that software failures cause the majority of recalls. The issues are system compatibility (interoperability between treatment planning and treatment delivery systems), user interfaces (human factors), and dose calculation (clinical decision support software). Similar to hardware, software also has hazards (any source of harm) and hazardous situations that can transform a hazard into harm. We must identify hazards, identify what event can turn them into harm, and write software specifications to prevent such combinations. Risk management tools such as Failure Mode and Effects Analysis (FEMA) and Fault Tree Analysis (FTA) can help to mitigate software risks. These tools are covered in Chapter 4. The mitigations from using the tools become the new requirements in the software specifications.

Negative Requirements Analysis for Worst-Case Scenarios

We often design for foreseeable circumstances, but we must be prepared for unforeseeable circumstances and worst-case scenarios. Instead of thinking of how we can make each function of the product work better, we can assume that each function can fail unexpectedly. Identify the high-risk failures and make sure the specifications include solutions for preventing selected failures that can result in harm.

Conducting PHA to Assess Risks

After the specifications are approved, it is time to confirm that no other harm scenarios exist in the product. This is done by a cross-functional team using the evidence-based structured method called Preliminary Hazard Analysis (PHA). It is briefly described in the ISO 14971 standard. As mentioned, the details are covered in Chapter 3 and also in Raheja and Allocco [6]. It is documented in the form of a table that contains the following:

■ Identification of the potential hazards in design, use, maintenance, repair, and product disposal by thinking of all the possible harm scenarios including sentinel, adverse, and never events
■ Results of risk assessment based on the probability and the severity of harm
■ Identification of mitigation strategies that must be added to the specification

Considerations for *In Vitro* Devices

An *in vitro* device (IVD) examines specimens from the human body to provide information for diagnostic, monitoring, or compatibility purposes. This includes reagents, calibrators, control materials, specimen receptacles, software, and related instruments.

Traditionally, IVDs have been used primarily by hospitals, clinical laboratories, and physicians' offices. But some devices are designed for home use, like blood sugar monitors, where no physician is involved to interpret the test results. For such devices, labeling must be simple, concise, easy to understand, make liberal use of illustrations and drawings, use bold print or other methods to highlight warnings and precautions, and provide color coding of reagent containers whenever possible. 21 CFR 809.10 covers labeling for home and professional use [7].

The labeling requirements should be included in the specifications. It should not be left to the discretion of only the marketing department. A legal review would be helpful from the product liability point of view. The label should not fall off the device because of an unreliable adhesive. Reliability should be included in the labeling material specifications. Devices for laypersons should be designed with a view to ensuring that the device's performance will not be significantly affected by anticipated variation in the user interface.

Summary

Poor specifications are the biggest reason for recalls. Don't be afraid to spend a lot of time on brainstorming using the tools described in this chapter. Doing right things right the first time is much cheaper than fixing too many flaws in production.

References

1. Newsmax. Not Engaged at Work? Gallup Poll Finds 70 Percent of Americans Aren't, June 25, 2013, Accessed on March 20, 2014 at http://www.newsmax.com/TheWire/not-engaged-work-americans/2013/06/25/id/511836/
2. Cause International website, *E-Waste Facts*. Accessed on March 21, 2014 at https://www.causesinternational.com/ewaste/e-waste-facts
3. Raheja, Dev, and Escano, Maria C. "Usability Hazard Analysis: Risk Reduction and Applications in Medical Product Safety Design and Usability," *Journal of System Safety*, March–April 2010.

4. U.S. FDA Class 2 Device Recall ONCOR Expression, November 6, 2007, http://www.accessdata.fda.gov/scripts/cdrh/cfdocs/cfres/res.cfm?id=65672

5. The FDA web site, Medical Device Recall Report FY2003 to FY2012, http://www.fda.gov/downloads/aboutfda/centersoffices/officeofmedicalproductsandtobacco/cdrh/cdrhtransparency/ucm388442.pdf

6. Raheja, Dev, and Allocco, Michael. *Assurance Technologies Principles and Practices*. New York: Wiley 2006.

7. U.S. Food and Drug Administration (FDA), "Assessing the Safety and Effectiveness of Home-Use *In Vitro* Diagnostic Devices (IVDs): Draft Points to Consider Regarding Labeling and Premarket Submissions," October 1988, http://www.fda.gov/MedicalDevices/DeviceRegulationandGuidance/GuidanceDocuments/ucm094277.htm

Chapter 3

Risk Assessment and Risk Management

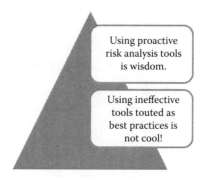

Using proactive risk analysis tools is wisdom.

Using ineffective tools touted as best practices is not cool!

Introduction

Specifications are usually incomplete, even after the attention described in Chapter 2 is paid to them. By the time you apply the information in Chapter 2, you have only the concept design. The next step is to conduct risk analysis to discover new safety-related risks, and therefore add more new requirements. The main tool for risk analysis is the Preliminary Hazard Analysis (PHA). We will go deeper into this most important analysis in this chapter. The other tools for deeper analysis are Failure Mode and Effects Analysis (FMEA) and Fault Tree Analysis (FTA). These are covered in Chapter 4.

Developing Risk Acceptance Criteria

Before any analysis is done, management should establish ranges of risks that are acceptable and unacceptable, which risks require further investigation, and those that require management review. Management should also make sure the definitions of risk and its components (the hazard and the hazardous situation) are clearly defined in the engineering standards and understood by the entire chain of command. This understanding is critical for clear communications.

Risk is defined as the combined impact of the severity of harm and the probability of harm. Therefore, we need acceptance criteria for each combination of severity and the probability of harm. This is done by developing a *hazard assessment matrix*, also called a *risk assessment matrix*. An example is shown in Figure 3.1. Each cell represents a combination of severity and probability of harm and is usually color coded to show if the risk is acceptable, unacceptable, or needs further investigation or management review. Red is

Frequency of Occurance	**Hazard Categories**			
	1 Catastrophic	2 Critical	3 Marginal	4 Negligible
(A) Frequent	1A	2A	3A	4A
(B) Probable	1B	2B	3B	4B
(C) Occasional	1C	2C	3C	4C
(D) Remote	1D	2D	3D	4D
(E) Improbable	1E	2E	3E	4E

Unacceptable risks: 1A, 1B, 1C, 2A, 2B, 3A
Unacceptable (investigate risk reduction):   1D, 2C, 2D, 3B, 3C
Acceptable with management review: 1E, 2E, 3D, 3E, 4A, 4B
Acceptable risk without approval: 4C, 4D, 4E

Figure 3.1   Example of a risk assessment matrix.

usually used for unacceptable risks, green for acceptable risks, and yellow where management review is required. There are more matrix options available in International Organization for Standardization (ISO) 14971 described in Chapter 1.

Risk Analysis Using PHA

Before we approve the specifications, we must brainstorm for all possible ways a patient can be harmed by the medical device and mitigate all major risks by designing them out. Fortunately, a technique—widely used in aerospace, aviation, and commercial transportation industries for decades—exists. This technique is called Preliminary Hazard Analysis (PHA). It covers all types of unsafe events, including sentinel, adverse, and never events. This is the first risk analysis technique in the ISO 14971 standard.

The Joint Commission on Accreditation of Healthcare Organizations (JCAHO) defines a *sentinel event* as an unexpected occurrence involving death or serious physical or psychological injury, or the risk thereof [1], such as harm caused by a malfunctioning surgical robot. Serious injury specifically includes loss of limb or function. The JCAHO defines an *adverse event* as an untoward, undesirable, and usually unanticipated event, such as the death of a patient, an employee, or a visitor in a healthcare organization. Harmful incidents such as those resulting from improper use of an infusion pump, or from its improper programming by the caregiver, or from an improper design of the software in the medical device, are also considered adverse events even if there is no permanent harm to the patient. The National Quality Forum (NQF) defines *never events* as errors in medical care that are clearly identifiable, preventable, and serious in their consequences for patients, and those that indicate a real problem in the safety and credibility of healthcare. Examples of never events include surgery on the wrong body part, a foreign body left in a patient after surgery, a mismatched blood transfusion, picking a wrong dose from the drop-down menu on the computer, a severe pressure ulcer acquired in the hospital, and preventable postoperative deaths.

The PHA process consists of the following:

- Identifying the potential hazards in the device, its use, and foreseeable misuse
- Assigning the criticality based on the probability of the harm and severity for each hazard
- Performing risk assessment
- Identifying mitigation strategies

This information is documented in the form of the matrix shown in Figure 3.2.

To start, we need to do some research on adverse, sentinel, and never events that are already known in the industry. Such data can be found on the websites of the Joint Commission, the Agency for Healthcare Research and Quality (AHRQ), the Leapfrog Group, the U.S. Food and Drug Administration (FDA), and Consumers Union. The Joint Commission, for example, includes more than 800 sentinel events on its website. Consolidate the data relevant to your product and compile a list of hazards that lead to such events. Remember that your device may work perfectly, but if it causes hazards in connected devices, then it can be a hazard related to your device. Add hazards from your own organization's experience, lessons learned, incident reports, and complaints from the families of patients. Such a list is called a *preliminary hazard list*. It is just a master list of hazards. In addition, cross-functional brainstorming should be done to add hazards that could occur but are not on the list. The next step is to assess the risk for each hazard in terms of the severity and the probability of the risk. The final step is to specify mitigation actions.

Preliminary Hazard Analysis

System:__Accucheck_Meter_____ Date_____ Team Members:

Item Number	System Component	Hazard Description	Hazard Effect	Severity/ Probability	Mitigation Strategy
1	Test strip	Customer using expired strip	Erroneous sugar readings that may result in automotive accidents.	I/D	Ask supplier to give warning to users in bold print on the container.
2	Glucose meter	False positive— sugar too high	User may take insufficient insulin, go into coma.	I/C	Design the meter for higher accuracy.
		False negative— sugar too low	User may take too much insulin.	II/C	Design the meter for high accuracy or warn users to make frequent checks.
3	Control solution to verify the meter accuracy	User ignores need to verify accuracy	User may take too much medication or too little medication.	I/B	Current instruction sheet is inside the box. Many users find it inconvenient to read. Put the instructions on the box.

Note: Severity Legend: I-Catastrophic, II-Major, III- Minor, IV- Negligible; Probability Legend: A-Very High, B-Moderate, C-Low, D-Very Low, E- Remote.

Figure 3.2   Example of the Preliminary Hazard Analysis.

Assessing the Risk

Risk assessment does not have to be a statistical process; it can be done qualitatively. Risk is the function of the severity of an event and the probability of harm. In the PHA in Figure 3.2, the qualitative values are used for both. The fifth column in this figure shows roman numerals for severity rating and alphabetical descriptions for probability ratings (A for the highest probability, and E for the lowest probability). There are four qualitative ratings for severity (some companies have three or five depending on their choice):

I—Catastrophic Harm: Potential for death
II—Major Harm: Potential for disabling harm or serious long-term illness
III—Marginal Harm: Requires medical intervention for a short time
IV—Negligible Harm: Does not require medical intervention

The probability scale is A through E. The meaning for each should be determined by each company. Examples are as follows:

A. Frequent: Can happen once a month or more frequently
B. Probable: Can happen once a year
C. Occasional: Can happen once in three years
D. Remote: Can happen once in five years
E. Improbable: Should never happen, but is possible

If we look at the first row in Figure 3.2, the consequence of an erroneous reading of sugar level can cause a fatality if the user drives a car unsafely with a high sugar level in the blood. Therefore, the severity rating is I. The chance of such a harm happening may be about once in five years, resulting in the probability rating of D. The combination of severity and probability is called *criticality*, which is I/D for this hazard. In other words, there is a remote chance that a fatality can happen. This is the statement of risk assessment.

Mitigating Risks Using World-Class Practices

The ISO 31000 standard is the generic risk treatment standard at the business level, and suggests selection of one or more of the following options for managing risks, and implementing those options. Risk treatment options are not necessarily mutually exclusive. The options can include combinations of the following:

Risk Transfer: Sharing the risk with another party or parties (including suppliers and insurers)

Risk Avoidance: Avoiding the risk by deciding not to start or continue with the activity that gives rise to the risk or removing the risk source

Risk Mitigation:
- Changing the likelihood
- Changing the consequences (impact)

Accept the Risk:
- Retaining the risk by informed decision
- Taking the risk to pursue an opportunity

Selecting the most appropriate risk treatment option involves balancing the costs and efforts of implementation against the benefits derived, with regard to legal, regulatory, and other requirements such as social responsibility and protection of the natural environment.

A number of treatment options can be considered and applied either individually or in combination. Risk treatment itself can introduce new risks. A significant risk can be the failure or ineffectiveness of the risk treatment measures. Therefore, monitoring needs to be an integral part of the risk treatment plan to give assurance that the measures remain effective.

At a technical level, the choices can be as follows:

A. **Mitigate:** Corrective action to eliminate or reduce the impact or likelihood
B. **Avoid:** Use different approach or different technology to avoid the risk

C. **Transfer:** Shift impact to another entity such as the suppliers

D. **Accept:** No corrective action feasible; document acceptance decision and monitor

ISO 14971 offers another alternative. It gives three options in the order of precedence:

- *Build inherent safety* into the design to eliminate the hazards.
- *Provide protective safety* in the form of safeguards such as failing in a safe default state, alarms, and redundancy.
- *Provide descriptive safety* such as warnings, highlighted in the equipment manual, highlighting important preventive maintenance, training, personal protective equipment, and visual aids.

Risk Evaluation

The purpose of risk evaluation is to assist in making decisions when the outcomes of risk analysis are questionable and no more mitigation seems feasible. In some cases, the risk is not acceptable but the benefits to the public outweigh the risks. Decisions should be made considering legal, regulatory, and other requirements. Of course, the FDA must approve the device.

In some circumstances, the risk evaluation can lead to a decision to undertake further analysis. There is always a better way.

Managing Residual Risks

After we do everything right, customers are likely to discover new risks. These were either overlooked or the operating environment was not fully understood. Typical sources of this information are as follows:

- Nonconforming reports on suppliers
- Nonconforming reports in production
- Incidence reports in hospitals
- Clinical data
- Postmarket surveys
- Complaints
- Medical device records
- Technical conferences

Summary

For a good risk assessment, we need to use at least three tools: PHA, FMEA, and FTA. Managing risks can be done in several ways as described in this chapter. The best strategy is to design out the major risks if possible by using different design approaches or different technology.

Reference

1. Raheja, Dev. "System Safety in Healthcare: Preliminary Hazard Analysis for Minimizing Sentinel, Adverse and Never Events," *Journal of System Safety*, July–August, 2009.

Chapter 4

Preventing Recalls during Early Design

Use Fault Tree Analysis to approve the new devices.

They help solve complex problems and lead to elegant solutions!

Introduction

Once the Preliminary Hazard Analysis (PHA) is done, the design concept for safety becomes clear. It is time to assess the risks at the detailed design level for reliability and durability. That is where the Failure Mode and Effects Analysis (FMEA) and the Fault Tree Analysis (FTA) come in. They help in refining the design using the old saying "the devil is in the details," and in developing the design structure for hardware and software. The structure can be expressed and communicated in the form of a functional block diagram. Some functions

are performed sequentially, while others are performed con-
currently. The functional block diagram shows the connections among different functions. This information is critical to
ensure that no new risks are created during design changes.
The purpose of the functional diagram is to communicate the
design functions and interfaces clearly to the entire product
development team during design reviews. Preventing recalls is
more effective if the criteria for selecting the right team members are given priority. The best performing teams include an
outsider as an independent voice. To start with, a team charter
should define the mission of the team and its objectives. The
charter should include a statement of work, the background of
the project, the thoroughly reviewed specifications, and should
define the extent of the team's authority, the accountability of
each member, the boundary conditions for analysis (how far
and deep you want to go), and responsibilities for interactions
with other groups and leaders.

Functional FMEA on Design Concept to Prevent Failures

Good design practices should consider all the design analysis
tools. Try to include these practices in your design standards.
After the tools are used, you may have to revise the design to
accommodate fault tolerance, fail-safe features, built-in self-
checks, user alerts, testability, user friendliness, and so on.

FMEA is a method for assessing the potential failures,
their failure mechanisms, and deciding the mitigation for the
device. It is based on an understanding of the relationships
between product requirements and the physical characteristics
of the product (including variation in the production process),
and the interactions of users with the product. FEMA takes
into account life-cycle knowledge of environmental and operating conditions and the duration of the intended applications.

To ascertain the risk of failure, a risk priority number (RPN), called *criticality*, is calculated for each failure mode. The RPN is the multiplication of the severity, occurrence over a defined time, and the detection rating, each usually on a scale of 1 to 10. The detection rating concerns how early the defect is detected. If the problem cannot be detected until the product fails, a rating of 10 is assigned. The higher the RPN, the higher the risk. Some FMEAs use only severity and probability of failure qualitatively to describe criticality, just like the ratings discussed in Chapter 3. FMEAs may use roman numerals, such as I, II, III, IV, V, for severity rating, and A, B, C, D, E for probability, as seen in Figure 4.1. Figure 4.2 shows criticality based on a scale of 1 to 10. FMEA can be conducted at several levels: system functions, subsystem, component, production process, customer use, or maintenance.

To perform a functional FMEA, we need the system specification, which is a compilation of all the functions. Each function is analyzed for its failure modes. A failure mode is the answer to the question: "What can go wrong in carrying out this function?" Some failure modes will be obvious, such as "the infusion pump may perform an unintended shutdown," or "the function may not execute at the right time."

The FMEA in Figure 4.1 includes a new column that is not typical in a FMEA—called Interface Effects. This column addresses how a failure of a function affects software, interacting devices, and human interfaces.

Conducting a Component-Level FMEA to Identify Parameters Critical to Quality

It is always a good practice to conduct a FMEA from different angles. The functional FMEA is done to assess system failures based on the failures of each function. We also need to look at the functions of the components to assess how malfunctioning

Product ___Spinal Cord Stimulator___   **Date**___

Specification Reference	Function	Failure Modes	Causes	Effects	Criticality	Robustness Actions	Interface Effects
0016	Deliver electrical simulation over the dorsal column of the spinal cord.	Excessive stimulation	The control device could give wrong stimulation.	Possible trauma to the patient.	I D	Develop statistical algorithm to limit stimulation based on the patient's history.	Design fail-safe limits in the control device in case of device malfunction.
		No stimulation	Electrodes not placed properly in the epidural area.	Patient unable to control pain.	III C	Warn the caregiver about improper placement of electrodes.	Caregiver training required.

Figure 4.1 **Example of a function analyzed using the functional FMEA.**

Component	Potential Failure Mode	Cause of Failure	Effects	Criticality				Recommended Action	Revised Risk			
				S	F	D	RPN		S	F	D	RPN
Lead wire	Wire can break	Weak material	Possible heart failure for the patient	10	2	10	200	Include alarm before the joint completely breaks.	10	2	4	80
		Weak joint to the heart valve	Possible heart failure	10	3	10	300	Design for high strength and longer life. Design redundant joint and warn when first wire breaks	10	1	1	10

Figure 4.2 **Example of a FMEA for components.**

components cause system failures. This FMEA also identifies which components are critical (those with a high RPN number) for safety and reliability. Once we have the list of critical components the FMEA should address the features of components that are critical to quality (CTQ) so we can identify the points in the manufacturing process that are critical. This information should be recorded in the Recommended Action column. This is the intent of the risk analysis tool Hazard Analysis for Critical Control Points (HACCP) in International Organization for Standardization (ISO) 14971 covered in Chapter 1. Figure 4.2 shows the FMEA for a pacemaker component.

The FMEA in Figure 4.2, as we mentioned, identifies CTQs also. In this case the strength of the joint to the heart valve and the inherent strength of the wire are two CTQs (the strength of the joint is improved with a redundant joint). Their performance must be validated.

Conduct an FTA to Develop Robust Solutions for Complex Problems

Dr. Mark Chassin, president of the JCAHO, says that the current method of root cause analysis, which assumes that a single event can cause an adverse event, is a misnomer because there is never a single cause in an adverse event [1]. The FTA highlights multiple combinations that usually cause harm. This is a *real* root cause analysis tool that must be used for medical devices.

FTA is a deductive process that can be especially useful for analyzing potential adverse and never events. Since many risks cannot always be foreseen by design, FTA is a powerful tool for identifying potential user mistakes, user misuse, maintenance mistakes, and software faults. It is an advanced root cause analysis tool structured as a visual cause-and-effect diagram. It uses standard symbols developed during the

Minuteman missile program. It differs from fishbone diagrams because it shows logical relationships and interconnections among causes. The purpose is to identify the initiators and contributors to these potential mishaps, thereby eliminating as many causes as possible in design and including installation of appropriate safeguards. FTA can also be used for analyzing complex design problems and building safeguards into the device design.

Complex examples where FTA can be used are as follows:

■ *Inherent dangers in technologies.* Radiation from computed tomography (CT) scans done in 2007 was estimated to cause 29,000 cases of cancer and kill nearly 15,000 Americans, according to findings published in the *Archives of Internal Medicine* [2]. Rita Redberg, a cardiologist and an editor for the *Archives* had this to say: "What we learned is there is a significant amount of radiation with these CT scans, more than what we thought."

■ *Software in the equipment dangers.* In 2005, a Florida hospital disclosed that 77 brain cancer patients had received 50% more radiation than prescribed because one of the most powerful—and supposedly precise—linear accelerators had been programmed incorrectly for nearly a year [3].

■ *User environment dangers.* A customer wearing a pacemaker died when he tried to go through the cashier in a department store. A magnetic device near the cash register that was supposed to detect stolen merchandise reacted with the software in the pacemaker, which resulted in the death [4].

The usual starting point for an FTA is the identification of a potential or existing mishap. The symbols used in aerospace designs are now standard symbols used in practically all industries including nuclear, chemical, automotive, and medical device. The symbols that are frequently used in a fault tree are shown and explained in Figure 4.3.

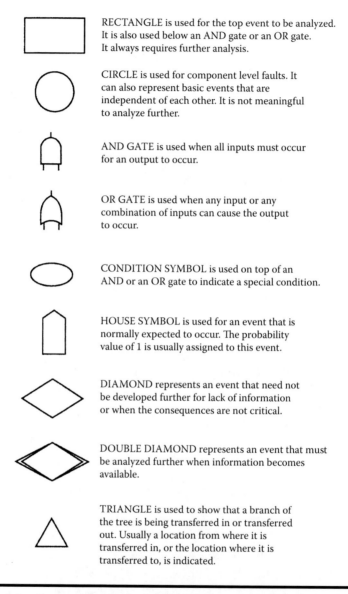

RECTANGLE is used for the top event to be analyzed. It is also used below an AND gate or an OR gate. It always requires further analysis.

CIRCLE is used for component level faults. It can also represent basic events that are independent of each other. It is not meaningful to analyze further.

AND GATE is used when all inputs must occur for an output to occur.

OR GATE is used when any input or any combination of inputs can cause the output to occur.

CONDITION SYMBOL is used on top of an AND or an OR gate to indicate a special condition.

HOUSE SYMBOL is used for an event that is normally expected to occur. The probability value of 1 is usually assigned to this event.

DIAMOND represents an event that need not be developed further for lack of information or when the consequences are not critical.

DOUBLE DIAMOND represents an event that must be analyzed further when information becomes available.

TRIANGLE is used to show that a branch of the tree is being transferred in or transferred out. Usually a location from where it is transferred in, or the location where it is transferred to, is indicated.

Figure 4.3   Frequently used standard fault tree symbols.

Let us begin with a mishap for a patient who is using a mechanical ventilator but receiving insufficient breathing assistance, which results in hypoventilation (low respiration rate). The top level of the tree is shown in Figure 4.4. Note that each of the five rectangles has a triangle attached. Each triangle

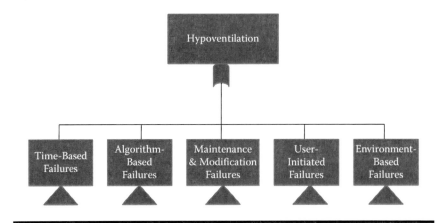

Figure 4.4   Top-level tree showing an OR gate relationship.

points to a different tree whose location, such as the page number, is identified inside the triangle (locations not shown here). This top level shows that the cause of hypoventilation can be due to any of the failures shown below the OR Gate. The OR gate symbol implies that any of the five events below it can cause the mishap. They can be time-based failures, or algorithm-based failures, or maintenance and modification failures, or user-initiated failures, or environment-based failures. Any one of these or any combination thereof can trigger hypoventilation. But keep in mind that the single event alone cannot cause a mishap. It is usually a result of a combination other hazards in a system shown in the trees represented by the triangles. The key to a safe device is to design out these hazards if possible or implement safeguards or redundancy.

Figure 4.5 shows an example of the next level of the tree for the fourth rectangle in Figure 4.4, *User-Initiated Failures*. Note that there are two AND gates in the diagram. Each of them implies that multiple things have to go wrong simultaneously before a user-initiated mishap takes place. Let us look at the rectangle labeled "user programming errors." It suggests that two causes must occur simultaneously before the programming error turns into a user-initiated failure: the respiratory therapist chooses a wrong control (controls volume

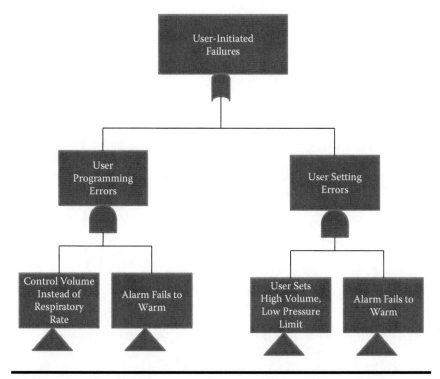

Figure 4.5 Fault tree showing the causes of the user-initiated failures.

instead of respiratory rate) AND the alarm fails to warn. If we continue building the fault tree, there may be six, seven, or more levels. Usually five levels are sufficient for developing solutions that mitigate risks.

Strategy for Developing Solutions

To develop a risk-mitigation plan, we need to know how many paths through AND and OR gates lead to the top mishap event. Whenever there is an AND gate, it means there is a safeguard or redundancy. The only way for the mishap to happen is if all the inputs to the AND gate fail. The system must be designed so that at least one of the inputs to the AND gate does not fail. Therefore, an excellent design strategy is to make sure the top level in the FTA contains an AND gate, not an OR gate. This ensures that the device cannot fail from a single failure. The only way the device can fail is when all the inputs to the AND gate fail.

Summary

The FMEA and FTA are the best tools to prevent device failures. For all safety-related mishaps, the design should include protection from failure represented by the AND gate in the fault tree. If there is an AND gate at the top, the risk of recall is drastically reduced!

References

1. Agency for Healthcare Research and Quality (AHRQ), *Web M&M-In Conversation with Mark Chassin, MD, MPP, MPH,* April 2009, http://www.webmm.ahrq.gov/perspective. aspx?perspectiveID=73
2. "FDA and Medical Device Recalls," FreeAdvice, http://injury-law.freeadvice.com/defective_products/medical-device-recalls .htm
3. Bogdanich, W. "Radiation Offers New Cures, and Ways to Do Harm." *New York Times*, January 23, 2010, http://www.nytimes .com/2010/01/24/health/24radiation.html
4. Raheja, Dev, and Allocco, Michael. *Assurance Technologies Principles and Practices.* New York: Wiley, 2006.

Chapter 5

Preventing Recalls during the Detail Design Phase

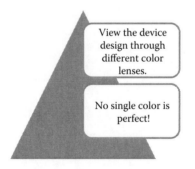

View the device design through different color lenses.

No single color is perfect!

Introduction

After making major design changes using Failure Mode and Effects Analyses (FMEAs) and Fault Tree Analyses (FTAs), we need to start designing for trustworthiness. This requires looking at different elements because each has a role in the safe operation of a device. Customers are looking for durability, reliability, inherent safety, inherent quality, forgiving design for user errors, minimum maintenance, trusted sterile packaging, process robustness, and safeguards from harm. If any of them is compromised, there can be serious consequences.

Designing for Durability

Seemingly harmless structural components, such as a humidi-fier lid hinge of a continuous positive airway pressure (CPAP) machine, might be strong enough to withstand a single applied load or a few open–close cycles. But what happens when the lid operates over and over, opening and closing, day after day, for years? Predicting component failure in such cases requires *fatigue* or *durability analysis*. Computer simula-tions can determine how well parts will hold up during cyclic loading. Results are important in calculating the minimum life and ensuring that the part meets the life-cycle duty require-ments. In case of critical care devices, such as heart implant-able devices that cannot be repaired easily, it is crucial to ensure that the moving parts that perform cyclic motions are extremely durable.

Durability analysis includes defect characterization, crack growth study, understanding the failure mechanism, and long-term performance prediction at different stress levels. In the past, durability analysis was largely the province of research. For starters, processes that cause part failure are complex. They involve progressive material changes, often highly localized, that end in cracks or complete fracture. For an introduction to durability analysis, it's helpful to start with a few basics. Since failure is progressive, knowledge of time or duty cycles to breakdown is an important element of design. And because failure is localized, it's necessary to understand local stresses.

In most cases, a constant load rarely causes a failure; the impact of fluctuating loads causes failures. When a crack gets larger, the stresses concentrate around the crack. The material experiences much higher loads and eventually breaks down. Failures, or fractures, take place when cracks get so large that the remaining material can no longer endure stresses and strains.

To design for durability, we need to recall our Engineering 101 classes, where we were asked to design with a 100% safety margin. This seems to be a forgotten art. Imagine a bridge designed for 20-ton trucks. It may have no problems in the beginning, but the bridge is degrading over time. After 5 years it may not be strong enough to take even 15 tons and it is very likely to collapse. If there is a stress concentration anywhere, it may collapse with even a 10-ton load. If the bridge was designed for 40 tons, it can be very safe even in the presence of a stress concentration. This is same as the 100% safety design margin we were taught in engineering schools. We can say that a minimum 100% safety margin compensates for uncertainties. For the same reason, electronic components in the aerospace industry and the medical industry are derated by 50% so that the actual load or stress is less than 50% of the strength. This helps to address deterioration of the components due to wear as well. We can also call this paradigm "Design for Twice the Load or Twice the Life." Fortunately, with a little creativity, designing for twice the strength can be very cheap. In a company in Michigan, design life was increased fourfold just by changing the heat-treating method on a shaft-and-key assembly. Similarly, the redesign of the first European jet aircraft (Comet) fuselage failures around the windows was done without increasing the thickness of the fuselage. They just changed the radius of the corners on the windows.

With a little thinking outside the box, one can see that designing for twice the life is cheaper than designing for one life. This requires an understanding of life-cycle costs. Usually, engineers talk of the average life when they estimate device life. This implies that 50% of the time the life will be less than the predicted life. In other words, either the device manufacturer or the customer has to pay for 50% failures during the product cycle. This is expensive for both. Besides, there are many indirect costs of monitoring, rework, and excessive inventory to replace failed parts. A company used the twice-the-life approach and turned the situation into a very positive cash flow. This is because there is nothing to be monitored, nothing to be repaired, and zero warranty cost since failures are going to occur beyond the first life. The 50% failure rate is now shifted to the second life, when the product is going to be obsolete [1]. Mold flow analysis, stress–strain analysis, dimensional stack-up analysis, fault tolerance analysis and derating analysis are some tools that can be used while designing for durability.

Once the analysis is complete, a durability test plan should be developed to test and assess the durability of the product. The durability test plan should be based on worst-case scenarios. Use–misuse scenarios should also be included in the test. These scenarios include events such as wrong control knob activation, lid opening–closing cycles, system in use dropped, and so on. The test plan should include testing for several lives. The following concept can be used to decide how many lives should be tested [2]:

One life, which represents nominal usage patterns of a typical user, where there is no safety issue.

Two times the life is a goal if there is any safety issue.

Three to four times the life to accommodate unexpected high loads and misuse scenarios.

Designing for Reliability

Critical failure is not an option in medical devices. When it comes to medical equipment, a device such as a patient-controlled analgesia (PCA) pump that works according to the specification but does not work well with interconnected devices can also pose the potential for harm. From critical care devices such as oxygen concentrators, lasers, ventilators, MRI scanners, insulin pumps, implantable pacemakers, to instruments as straightforward as stethoscopes, injection needles, and thermometers, medical devices must be reliable. Reliability is the ability of a product or a system to perform as intended without failure for a specified time, in its life-cycle environment.

The medical industry must have higher reliability standards than other industries. However, a high reliability standard is hard to maintain in today's environment of intense global competition where pressure for shorter product-cycle times, cost cutting, and higher customer expectations for quality and reliability are common. A good design for reliability (DFR) process will help ensure that the medical product is safe and reliable. For the entire process of DFR, see reference [3].

In a medical device, a critical failure is one that prevents or shuts off functions that are needed for medical interventions. Such failures can be life threatening for a patient, depending on the type of device. Since critical failure is not an option, it is worth repeating that FMEAs and FTAs should be performed to identify all failure mode scenarios. Make sure the usage scenarios identified in the use–misuse model are considered during the analysis. Once all failure modes are identified, select the failure modes that could cause critical failures and try to mitigate them through design, if possible. For example, if a joint on a pressurized container can leak because of excessive variation in its components, the failure can be prevented by choosing a jointless container design. If there are no joints, there will be no leaks.

If it is not possible or viable to design out the critical failure mode, the team must consider alternate design strategies, such as fault tolerance, in the form of active or passive redundancy. As an example, a medical lab device required cooling of thermal electronics with a fan. The failure of the fan could result in false positives in the test results. The problem was solved by using two lower-priced fans in active redundancy; the two low-capacity fans shared the cooling load, but if one of them failed, there was still sufficient cooling to prevent false positives.

In addition, the team must design a comprehensive test plan that will test all the known failure modes as well as all the potential failure scenarios that may have been overlooked. This is done by *fault insertion testing*, where various hardware and software faults are inserted to make sure the device behaves as designed and does not result in any unsafe behavior such as a false output or an unexpected malfunction.

A robust and reliable alarm system that annunciates a failure is a good backup in case of a failure. Always try to build redundancy into the fault-detection and annunciation (alarm) system. Try to source critical components in each redundant path from different vendors. This way, if a manufacturer ships a bad lot of critical components, both of the redundant paths will not have the same faulty component. Always perform shock and vibration testing, environmental testing, drop testing, and electrostatic discharge (ESD) testing on electrical devices, and fault-insertion testing on alarm systems. There are several safety standards recommended by the U.S. Food and Drug Administration (FDA) related to alarm systems. One example is the International Electrotechnical Commission (IEC) standard 60601-1-8:2006, Medical electrical equipment, Parts 1–8: General requirements for safety; Collateral standard: General requirements, tests, and guidance for alarm systems in medical electrical equipment and medical electrical systems (http://www.ansi.org).

Designing for Inherent Safety

Product safety is best characterized by the words *system safety*, which are used in the aerospace and defense industries. Some ISO 14971 tools are taken from this system thinking, especially the Preliminary Hazard Analysis (PHA). *System safety* is defined as the application of engineering and management principles, criteria, and techniques to optimize safety within the constraints of operational effectiveness, time, and cost throughout all phases of the system life cycle. This definition suggests that merely a good design is not the goal; rather, the goal is a successful design, one that anticipates problems in advance, is implemented on time and minimizes life-cycle costs.

Often tests are not done under conditions that duplicate field environments, and many human errors, such as making faulty wiring connections and shipping parts with wrong labels, are easy to make. Device designers should not wait for reports on accidents from these mistakes. This concern for preventing accidents without the help of much data led to broadening the concept of *product safety* to *system safety*. It is the safety of the system that is important, not just a safe product.

The concern for probability uncertainty of the system failing (such as false alarms, false negatives) becomes more of an issue in the healthcare industry. To understand system safety, one must understand the definition of a system as "a composite, at any level of complexity, of personnel, procedures, materials, tools, equipment, facilities, environment and software" [4]. The elements of this composite entity are used together in the operational or support environment to perform a given task or achieve a specific mission. In the healthcare industry, we must expand the definition of a system to include patient handoffs, verbal communication, and interface between functional departments.

The steps in the process are as follows:

1. Identify the risks using hazard analysis techniques such as PHA and FTA as early as possible in the system life cycle.
2. Develop options to eliminate, control, or avoid hazards.
3. Provide for timely resolution of hazards.
4. Implement the best strategy.
5. Control the hazards through a closed-loop system.

Note in item 2 above that the system safety approach does not rely on warnings and instructions in manuals. These are important to include, but they don't prevent all accidents. As long as humans are involved, some will ignore instructions and warnings. Does everyone on a highway stay under the posted speed limit?

Designing for Inherent Quality

Designing for quality means designing the product in such a way that it does not produce manufacturing defects.

Examples include the following:

■ If two components, when assembled, can result in a defect, design them as a single piece.
■ If a fastener can come loose over time, design a clip joint or a locking joint.
■ If a soldered wire to a heart valve can break because of poor soldering, use a redundant wire.
■ If a component can be assembled backward, design it so that it will not fit backward.

The belief that these actions may increase the cost of the product is a myth. The real cost usually goes down because of zero warranty costs, zero lawsuits, and zero recalls from quality defects! This is all explained in the famous book *Quality is Free* by Philip Crosby.

How do design engineers know these problems in advance? The answer is simple, but implementation is hard because of time pressures and lack of leadership. Quality problems can be predicted by conducting a process FMEA on the proposed process. The analysis requires participation by manufacturing engineers, design engineers, production supervisors, maintenance engineers, production operators, reliability engineers, and a quality assurance engineer.

Designing to Forgive User Errors

A patient was given a PCA pump to control the amount of pain killer depending on his needs. The patient delivered too much pain killer in order to reduce the pain quickly and went into respiratory depression. Too much medication blocked the passages to his heart. When he woke up, he saw eight doctors and caregivers administering emergency interventions. This mishap could have been avoided if the device could not harm him in spite his error. About 13 million patients use this device each year [5]. As discussed earlier, the device could have been programmed to limit the size of the dose, regardless of the patient's choice but sufficient to keep the patient safe. This is an example of designing to forgive the user for errors. Medical device manufacturers are developing smarter alarm systems that can indicate when a patient is trending into unsafe territory [6].

Again, how do design engineers know these problems in advance? The answer is the same as in designing for quality. Conduct a user FMEA for the hospital use process. Designers need to have the correct participants: the nurses, doctors if possible, a hospital quality assurance specialist, and patient safety officers who have used the device in the past.

Designing for Hazard-Free Maintenance

An anesthesiologist in a hospital wanted to deliver oxygen to a patient prior to surgery. Instead, he put the patient in a coma because an incorrect gas was delivered. The reason was that the maintenance technician connected the oxygen hose to the wrong gas supply. Hospitals are now willing to pay more for a device in which each hose has a noninterchangeable fitting.

The same question arises again. How do design engineers know about maintenance errors in advance? The answer is still the same. Conduct an FMEA on the maintenance process. Team members include maintenance technicians, maintenance engineers, and users of the equipment.

Designing for Packaging

There have been many recalls concerning packaging: wrong components, wrong labeling, unsterilized parts, and contaminated bags and poorly sealed bags. The major reason is the increase in more complex products [7]. These include combination products (such as the pairing of a delivery device and a pharmaceutical), the growth of kits, larger products made up of more items connected together to reduce the amount of manipulation and handling required by caregivers, and more microbiologic-centric products such as skin tissues. These issues must be considered during package design.

Protection of the product is a major issue in packaging design. It must withstand shipping and handling damage, especially in overseas shipments where environmental extremes and roughness in handling may be unknown. Packages can sit in high temperatures and rain for months in open storage in warmer countries. The design specification must include worst-case scenarios and shelf life. Accelerated-aging testing is critical to understanding potential package weaknesses and knowing how to address product damage problems that sometimes occur even years after the product has been on the market [7]. Companies should monitor their products for years for new packaging risks. A test guide, such as ASTM F1980, Standard Guide for Accelerated Aging of Sterile Medical Device Packages, can be used to assess some designs.

Many packaging problems can be avoided if the knowledge gained by others is shared. For sterility issues, the study by Hall can help [8]:

> The underlying cause of defective heat seals was examined in packages produced at various sites in many countries. In more than 90% of these incidents, inadequate contact pressure at the seal site led to

failure. For the others, inadequate heating of the mating surfaces was the main issue.

When sterile barrier packs are found to have a poor heat seal this is naturally a major cause for concern. This defect can manifest itself in many forms. It can be a general reduction in seal quality throughout the seal area or the defect may be intermittent, and in other cases only part of the seal area is affected. The defect may be apparent to the eye when examining the unopened package, but often defects only become visible when the package is opened. However, new ultrasound scanning techniques now enable online detection of sealing defects and make possible early detection of problems.

Durability Testing

Knowing product durability is a part of the detail design before a product is transferred to manufacturing. This requires accelerated life testing. You must know the first failure mode. This can be found through clinical trials or by trying out the device in a simulated environment. The right procedure is to test samples at a minimum of three higher loads. The load can be heating–cooling cycles, random vibrations, wet–dry cycles for sensors, or a constant load. The selection of load depends on which load duplicates the field failure. In the example below, a constant high-voltage load was able to duplicate the field failure.

The next step is to determine the minimum life for each sample (many engineers measure the average life, but that comes with 50% failures at that life) and plot the minimum life on log-log graph paper. For safety-related failures, it is prudent to predict minimum life instead of the average life. These three points will follow a straight line if the test is duplicating

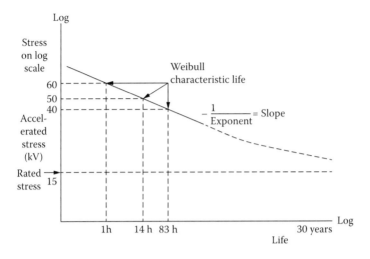

Figure 5.1 **The S-N diagram for life testing. (From Dev Raheja and Michael Allocco, *Assurance Technologies Principles and Practices*. New York: Wiley, 2006).**

the failure discovered during the clinical trial at normal load. If the failure mode on the test is different, we don't have the right test. This diagram is called the S-N diagram (stress versus the number of cycles/hours/months diagram). See Figure 5.1 for an example.

To predict minimum life at normal stresses, project the graph line to the normal stress level as shown in Figure 5.1. This relationship can be mathematically described by the equation below, where N is the reciprocal of the slope of the S-N graph line. Once we have the model, we can determine the accelerated test time at any chosen load.

Example: A high-voltage product rated at 15 KV failed frequently. The company had lost more than 60% of its customers. The company changed the design so that no failure would occur for 15 years. Since the use life is targeted for 15 years, they tested for 30 years of use (100% safety margin, design for twice the life). Since the test time at higher loads is measured in hours, we need to convert 30 years into hours (30 × 365 day × 24 hours per day) in order to use the equation. The company was able to re-create the field failure mode at four times the load (60 KV) to reduce the test time. The reciprocal of the slope was 8.2. The only unknown in this equation was the test time (life at accelerated load). This was calculated to be 3.04 hours as shown below. In other words, to prove that the device has a 30-year life, the test time is only 3.04 hours if the device can take a 400% load. This model applies to electrical, electronic, and mechanical devices. The model is called the *Inverse Power Law*. The equation is

$$\left[\frac{\text{Life at normal load}}{\text{Life at accelerated load}}\right] = \left[\frac{\text{Accelerated load}}{\text{Normal load}}\right]^{N}$$

where N is the reciprocal of the slope of the S-N plot, and is called the *acceleration factor.*

Let's use real numbers in this equation. The device is active 24 hours a day, every day for the entire life. The desired life was 30 years (30 years × 365 days × 24 hours). The normal load is 15 kilovolts; the accelerated load is 60 kilovolts (the failure mode at both the loads had to be identical). The accelerated time (the test time) works out to be

$$\left[\frac{30 \times 365 \times 24}{\text{Accelerated test time}}\right] = \left[\frac{60 \text{ KV}}{15 \text{ KV}}\right]^{8.2}$$

Accelerated test time = 3.04 hours

For fluids and chemical reactions, a different model, called the *Arrhenius model*, applies as follows:

$$\text{Life} = Ae - E/kT$$

where k is the Boltzman constant for the material, E is the activation energy (slope of life versus temperature plot on special Arrhenius graph paper), T is the temperature in degrees Kelvin, and A is a constant determined by plugging the life at any chosen temperature from the graph into this equation. The entire procedure for these two models can be found in reference [4].

Summary

The detail design must use different dimensions of customer perspectives. Designing for reliability, durability, safety, and other attributes covered in this chapter requires a separate approach for each.

References

1. Raheja, Dev. "System Safety Engineering." Tutorial presented at the Annual Reliability and Maintainability Conference, Orlando, FL, January, 2011.
2. Hegde, Vaishali, and Raheja, Dev. "Design for Reliability in Medical" Presentation at the Annual Reliability and Maintainability Conference, San Jose, CA, January, 2010.
3. Raheja, Dev, and Gullo, Louis. *Design for Reliability*. New York: Wiley, 2011.
4. Raheja, Dev, and Allocco, Michael. *Assurance Technologies Principles and Practices*. New York: Wiley, 2006.
5. Wong, Michael. "Managing Risk with Patient-Controlled Analgesia," Physician-Patient Alliance for Health & Safety, August 6, 2013, http://ppahs.files.wordpress.com/2013/08/michael-wong-article-mq-summer-2013.pdf
6. Jarzyna, Donna et al. "American Society for Pain Management Nursing Guidelines on Monitoring for Opioid-Induced Sedation and Respiratory Depression," *Pain Management Nursing* 12, no. 3 (September 2011): 118–145, http://www.aspmn.org/organization/documents/GuidelinesonMonitoringforOpioid-inducedSedationandRespiratoryDepression.pdf

7. Butschli, Jim. "Greater Product Complexity Demands More of Healthcare Packaging," *Packaging World*, March 13, 2013, http://www.packworld.com/applications/healthcare/greater-product-complexity-demands-more-healthcare-packaging
8. Hall, R.E. "Identifying Packaging Errors," China Medical Device Manufacturer, Fall 2008, http://archive.cmdm.com/article.php/ArticleID/2555?lang=en&

Chapter 6

Designing for Prognostics to Protect Patients

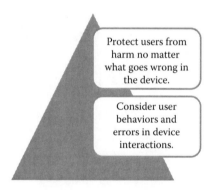

Protect users from harm no matter what goes wrong in the device.

Consider user behaviors and errors in device interactions.

Introduction

The role of prognostics in design is to protect patients by alerting caregivers and patients when equipment malfunctions or is about to fail. It is also an approach to protect the integrity of equipment and avoid unanticipated operational problems leading to mission performance deficiencies, degradation, and adverse effects on safety. Researchers have developed a variety of approaches, methods, and tools that are useful for these purposes, but applications in real-world situations may be hindered by a lack of real visibility into these tools, uniformity in application of these tools, and consistency in their demonstrated results.

There are at least four distinct requirements for product design:

1. The ability to prevent false positives and false negatives; alarms in devices are a big concern of the FDA.
2. Alerts when the device is not performing accurately, including identification of the problem.
3. Alerts the user when a device is near the low end of the prognostic distance (time to failure).
4. Shutdown of the device in a safe state if the failure cannot be avoided.

Preventing False Positives and False Negatives

False positives and false negatives are not limited to laboratory results, but also occur in medical devices such as MRIs, alarms on patient monitors, infusion pumps, and pacemakers. A woman did not know that she had a brain tumor. The hospital's MRI was not working accurately, but the radiologist did her MRI regardless, hoping that even an inaccurate MRI would indicate a problem. The tumor did not show up on the MRI. The doctor gave her an incorrect diagnosis, based on the false negative, with an incorrect prognosis. Her condition kept worsening and she became paralyzed on the right side of her body for life. She had to give up her job and is in a wheelchair for life.

False alarms, from the perspective of hospitals, are a big concern in intensive care units (ICUs) and in critical care. That is why the design of alarms on devices is also a big concern of the U.S. Food and Drug Administration (FDA). There are also numerous alarm-related hazards in medical devices used by patients. Such devices include intravenous pumps, feeding pumps, ventilators, cardiac monitors, compression devices, dialysis machines, nurse call systems, hospital beds, and chairs with exit alarms. Some additional devices in use with alarms include medication dispensing systems such as the Pyxis system, pocket phones, pagers, and telephones as secondary alarms. There are several factors that make it more difficult to hear alarms. These factors include the size and configuration of units, closed doors to patient rooms, and high noise levels on units. Mitigation of such risks by design is clearly needed. Some device alarms sound identical. For example, some patient beds and infusion pumps alarms are similar, making it difficult to distinguish which device alarm is sounding. Specifications should find creative ways to avoid harm.

At least 216 deaths nationwide between January 2005 and June 2010 were linked to alarms on patient monitors that tracked heart function, breathing, and other vital signs. In many cases, medical personnel did not react with urgency or did not notice the alarm, a type of desensitization that occurs from hearing alarms, many of them false. At least 119 people have died since 2005 in incidents involving warning alarms on ventilators, such as caregivers failing to respond to beeps warning of a problem or setting alarms improperly so they sounded too softly or not at all [1].

Alarm discriminability, the clinician's ability to distinguish one medical device alarm from another to respond correctly to the actual alarm, is a major concern of the FDA. Problems with alarm volume may occur when the alarm sound is not loud enough, can't be adjusted to be loud enough for the responder to hear, or when settings have been changed and not reassessed frequently enough. Issues with alarm activation thresholds also occur when the sensitivity levels for a given medical device alarm are adjusted based on a clinical situation or environment, left at that setting, and not readjusted for a new patient or new clinical situation.

From a device manufacturer's point of view, an alarm gives a true positive, if it works. From the point of view of hospitals, if the alarm does not represent the need for emergency intervention, then it is a false positive. Over 90% of alarms, even though required on devices, are false positives from the hospital's point of view. The Healthcare Technology Safety Institute of the Association for the Advancement of Medical Instrumentation (AAMI) was trying to establish a guide at the time of this writing. It works closely with the FDA on setting device standards.

What can designers do? There are actions that designers can take. Meet with doctors, nurses, and biomedical engineers to discuss the importance and priority of device alarms. If the alarm is critical, then discuss what sounds will distinguish it from noncritical alarms. They may require an audiovisual message at the nursing station in addition to the alarm, or require a message on mobile devices carried by nurses and doctors.

Designing for Alerts When the Device Is Not Performing Accurately

An FDA report shows that a balloon catheter failed to advance to the right position during cardiac surgery [2]. The balloon ruptured because of a scratch on the balloon. The balloon had partially opened because it had blood in it. The product was recalled. No mention was made about recall alerts to doctors, but the purpose of these alerts is to inform doctors before they start to use the balloon. Perhaps this could have been done by including a balloon test in the device, not just for balloon defects, but also for catheter leaks and other problems. Such problems can be foreseen by searching the FDA Manufacturer and User Facility Device Experience (MAUDE) reports, conducting a design review with doctors, or just interviewing doctors.

Designing to Alert When a Device Is Near the Low End of the Prognostic Distance

There were 121 peritoneal dialysis catheters placed in 81 children. Operative technique, catheter selection, and patient variables (e.g., age or prior surgical history) may influence catheter life span. The median functional catheter lifetime was 109 days. Thirty percent of the catheters failed within 60 days. Catheters placed without simultaneous omentectomy (surgical removal of the omentum, which is performed in cases where there may be spread of cancerous tissue into the omentum), were found more likely to fail [3]. The report concluded that omentectomy at the time of catheter placement decreased the risk of early catheter failure. This type of data gives designers clues as to when the device is likely to fail and allows monitoring of some variable that may indicate that the device is getting close to the end of its life. Monitoring the tire pressure in cars is an example of a warning before the tire fails.

Batteries in medical devices, such as pacemakers, infusion pumps, and defibrillators, are good candidates for designing for alerts. The amount of charge remaining is a good indicator of end of life for a battery. Many devices already do this.

Designers can anticipate the need for alerts by brainstorming with the hospital staff. Of course, designing for high reliability and durability is the best tool for avoiding the need for alerts in many cases.

Shutting Down the Device in a Safe State if the Failure Cannot Be Avoided

We discussed batteries, catheters, and other items that can fail suddenly because of many other causes. Regardless of how they fail, it is a good design practice to have the failure result in a safe default state if possible. In Chapter 4, a pacemaker example was discussed where a patient in a department store died because the pacemaker software setting was changed by the magnetic field in the cashier area, resulting in an excessive breathing rate. If the software had been designed to default to a safe breathing range, the mishap would have been averted. In the past, many pacemakers malfunctioned when patients drove near power company electrical towers.

Prognostics health management (PHM) related to batteries is particularly important since battery failures can result in disruption data failures. An FDA recall had the following conclusion [4]:

> When the GemStar Lithium battery voltage level drops below 2.4 volts, an "11/004" error is displayed and the device is rendered inoperable. This failure mode results in a delay/interruption of therapy. Additionally, infusion settings and event history logs will be erased as a result of this device malfunction. The severity of the clinical impact, due to the delay/interruption in therapy, is dependent upon the underlying condition of the patient and the treatment being prescribed. A delay/interruption in therapy has a worst case potential to result in a significant injury or death.

Progress in Prognostics Health Monitoring

The Institute of Electrical and Electronics Engineers (IEEE) was developing standard P1856, titled "Standard Framework for the Prognostics and Health Management of Electronic Systems," at the time of this writing. This standard aims to provide practitioners with information that will help them make business cases for PHM for electronic devices in general. Their PHM system performance includes metrics such as

- Failure detection accuracy (false positives and false negatives)
- Prognostic distance (time to failure)
- Prognostic accuracy (accuracy of prediction versus actual failure)
- Health management response time (time to react to predicted failures)

These metrics include the following in implementing the PHM:

- *Canary device.* This is a component that provides early warning of impending failure. This component should experience the same environmental and operational loading conditions as the system that it is monitoring. It should fail by the same failure mechanism as the monitored system.
- *Condition-based maintenance.* This is a preventive and predictive approach to maintenance based upon the evidence of need.
- *Corrective maintenance.* These are activities undertaken to isolate (diagnose) and rectify a fault so that a failed system can be restored to its normal operable state. Corrective maintenance can be either scheduled or unscheduled.
- *Failure threshold.* This is the acceptable limit(s) beyond which changes in the measured value(s) cause a function performance to become unacceptable.

Summary

Prognostic monitoring of a device is similar to monitoring the vital signs of a person: heart rate, respiration rate, blood pressure, oxygen level, and so on. By monitoring these variables, doctors initiate immediate action when an unsafe threshold is reached. Similarly, users need to be warned of necessary immediate actions when an unsafe performance threshold has been detected through monitored variables in a medical device.

References

1. Raheja, Dev, and Escano, Maria, C. "Hazards in Patient Monitoring Alarm Systems." *Journal of System Safety* 48, no. 3 (2012), http://www.system-safety.org/ejss/past/mayjune2012ejss/healthcare_p1.php
2. U.S. Food and Drug Administration (FDA). *MAUDE Adverse Event Report: Abbott Vascular-Cardiac Therapies Voyager Rx Coronary Dilatation Catheter,* 2009, http://www.accessdata.fda.gov/scripts/cdrh/cfdocs/cfmaude/detail.cfm?mdrfoi__id=1593298
3. Kribbs, R. K. et al. "Risk Factors for Early Peritoneal Dialysis Catheter Failure in Children." *Journal of Pediatric Surgery* 45, no. 3 (2010): 585–589, http://www.ncbi.nlm.nih.gov/pubmed/20223324
4. U.S. Food and Drug Administration (FDA). FDA Class I Recall, Hospira, Inc., GemStar Infusion System, Lithium Battery, Low Voltage, March 18, 2013, http://www.fda.gov/MedicalDevices/Safety/ListofRecalls/ucm349866.htm

Chapter 7

Preventing Recalls during Production Validation

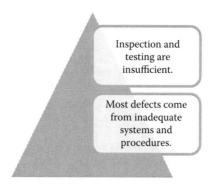

Inspection and testing are insufficient.

Most defects come from inadequate systems and procedures.

Introduction

Most manufacturing problems can be eliminated by better design of components and assemblies. One can call this approach *design for manufacturability*, which is the ability to produce the assembly without known defects, wherein very little or no inspection is required, there are no rejects, no rework, no warranty costs, and therefore high quality at low cost.

Understanding Key Design Features That Result in Defect-Free Production

The key to understanding production-related defects overlooked in design is the process Failure Mode and Effects Analysis (FMEA). In this FMEA we analyze what can go wrong in each step of the production process, and brainstorm with production employees to see if we can design out the production problems. This author's experience shows that employees can easily come up with design solutions 90% of the time. If unable to design out the problems, then manufacturing and quality assurance engineers should design an early warning system as a part of the process design or develop automated inspections. This may require a minor investment, but the benefits are many including no defects, no rework, no warranty costs, and increased customer satisfaction.

Examples of production-related defects are

- The guide wires designed to facilitate percutaneous coronary interventions for eight cardiac rhythm devices were recalled because the coating on their surface was delaminating and detaching. This Class I recall had a reasonable probability that use of, or exposure to, a defective product would cause serious adverse health consequences or death [1].
- The joy sticks on power wheelchair models did not work correctly. For details, see [2].

Process FMEA should be done before transferring a new design to production. Each process failure mode can mean one design change. It is prudent to make all the design changes at one time while the design is still on paper. Therefore, the process FMEA should be done immediately after the design FMEA. It is too expensive to find out that you have a wrong process when you have are already paid for it.

An example of designing out production defect: A plastic instrument panel was required to be conductive; therefore, it was coated with zinc plating. The production operators were unable to simultaneously control about 19 variables, all of which had an effect on the plating integrity. All 19 sources of defects were eliminated by mixing carbon in the plastic compound. Because carbon is a conductive material, the entire plating operation was eliminated, resulting in future savings of over $1 million. The cost of adding carbon was insignificant.

An example of designing an early warning in production: A thin sheet of steel required a very uniform thickness throughout the entire sheet. The process consisted of taking a very hot thick sheet and squeezing it between rollers at seven stations. One hundred percent inspection was required. The production employees suggested controlling the surface smoothness on the rollers and their eccentricity (out of roundness). They kept the data on eccentricity versus quality, and the surface smoothness of the rollers versus quality. They discovered that no defects were produced until the eccentricity exceeded an observed limit. They installed a sensor to warn when this limit was being approached. They changed the rollers before any defect was produced. They took a similar approach for the eccentricity (roundness) of the rollers. No more inspection of the sheets was required. No more defects! Millions of dollars were saved.

Understanding the Theory of Profound Knowledge for Superior Quality

No individual had more influence on quality management than Dr. W. Edwards Deming. Shortly after World War II, he was invited to help Japan. He defined a product or service as possessing quality if the product or service helps somebody and enjoys a good and sustainable market [3]. This definition is a win–win value proposition for the user as well as for the manufacturer, and it meets the intent of regulations.

He taught that management is responsible for 85% of the quality problems. He communicated to management saying that mistakes in production are a symptom of a troubled system, not the cause. Today we have competition from many countries so our ability to keep the company healthy depends on this knowledge. He implemented his doctrine with the so-called Theory of Profound Knowledge. It consists of four paradigms for management:

1. Appreciation for a system
2. Understanding of variation
3. Theory of knowledge
4. Understanding of worker psychology

Appreciation for a system implies that everyone should understand that a system is more than a sum of the parts. It includes suppliers, customers, end users, support staff such as maintenance technicians and cleaning crew, software running the process, facility design, and trustworthy communication among all in the system. That means working with suppliers and customers as partners in reviewing products and processes.

Understanding variation means knowing how much variability is inherent in the production process and how it comes from external causes called *special causes*. If the internal variation is high, it usually means the process is unacceptable. If the external causes are frequent, we must eliminate the source of the problem. For example, if the cause of variation is because maintenance in the production equipment was not done on time, then we can install reminder lights in the process and make the operator responsible for notifying the maintenance supervisor.

Theory of knowledge is about understanding the limits of current knowledge and of what can be known. There is very wise saying: If you know what you don't know, then you know!

Knowledge of psychology is about understanding the limits of human nature. Humans will make mistakes. It is only a matter of time. We must have a system that prevents defects from going to customers in spite of mistakes. We can use a poka-yoke (mistake-proofing) system to catch mistakes or reliability tests on production samples to catch hidden workmanship problems.

Conducting HAZOP Analysis to Identify Latent Hazards in the Manufacturing Process

A Hazard and Operability (HAZOP) (IEC 61882 and ISO 14971) analysis is based on a theory that assumes that process risk events are caused by deviations from the design or operating intentions. It originated in the chemical industry and has expanded to other industries. It is a systematic brainstorming technique for identifying hazards in potential deviations from normal use or design intentions [4]. HAZOP is similar to the design FMEA for deviation from the design intent and similar to the process FMEA for manufacturing deviations, but the *focus is on safety-related deviations only.* It uses a team of people with expertise covering the design of the manufacturing process and its applications. It can be applied to outsourced production as well as upstream suppliers, equipment, and facilities. The output of a HAZOP analysis is a list of critical operations for risk control. This facilitates regular monitoring of critical points in the manufacturing process.

Using ISO 14971 HACCP Analysis to Identify Critical Steps in a Process

According to the U.S. Food and Drug Administration (FDA) *Guidance for Industry, Q9 Quality Risk Management,* Hazard Analysis and Critical Control Points (HACCP) is a systematic, proactive, and preventive tool for assuring product quality, reliability, and safety. It is a structured approach that applies technical and scientific principles to analyze, evaluate, prevent, and control the risk or adverse consequence(s) of hazard(s) due to the design, development, production, and use of products [4].

It consists of the following seven steps:

1. Conduct a hazard analysis and identify preventive measures for each step of the process
2. Determine the critical control points
3. Establish critical limits
4. Establish a system to monitor the critical control points
5. Establish the corrective action to be taken when monitoring indicates that the critical control points are not in a state of control
6. Establish system to verify that the HACCP system is working effectively
7. Establish a record-keeping system

The technique can be used to identify and manage risks associated with physical, chemical, and biological hazards (including microbiological contamination). It is most useful when product and process understanding is sufficiently comprehensive to support identification of critical control points. The output of an HACCP analysis is risk management information that facilitates monitoring of critical points, not only in the manufacturing process, but also in other life-cycle phases.

Assuring Conformance to Key Design Features without 100% Inspection or Testing

Earlier in this chapter there was an example where a plating defect was no longer possible in a plastic panel, and no more inspection or testing was required for that defect. The message is that we should try not to depend on inspection or testing for known defects. Inspection is, at most, 80% effective according to quality gurus. This author's experience as a supervisor of inspectors and testers at GE Healthcare supports this finding. A good paradigm is "No inspection is the best inspection." This has to be seen in the right context, which is, design your process or product so that no inspection or testing is required. This is not always possible, but it can be done most of the time. If we cannot eliminate the defects permanently, then safety-related defects should be inspected either through sampling or through 100% inspection if sampling cannot ensure quality.

One more example may help. A process required two plastic components to be glued together. It was very hard to judge how long the glue could hold the pieces together in actual use. Such a process is called a *special process* in the FDA quality system regulation because the outcome is unpredictable. A company is required to develop special accelerated tests to prove the glue strength. This has to be proved not just during production validation, but also at periodic intervals, since processes and operators change over time. The need for inspection and testing was entirely eliminated by eliminating the joint and using a single-piece mold assembly.

Auditing to Identify Unacceptable Variation before Defects Are Produced

To prevent recalls, we need to conduct a special audit of safety-critical dimensions and other features (such as chemical properties, strength of solder joints, surface finish, and contamination) on new processes. Identification of safety-critical features should come from the Preliminary Hazard Analysis (PHA), FMEA, and Fault Tree Analysis (FTA). At a minimum, Six Sigma quality is required on these features. The Six Sigma concept requires that all parts must be within the middle half of engineering tolerances and that the process should able to maintain this level of quality over time over all three shifts with different operators.

If the process is not capable of maintaining the middle half of the tolerance, we don't have the right process. Then we should either change the component design or change the process.

Taking Corrective and Preventive Actions Using the FDA System

Corrective and preventive action (CAPA) is an FDA system for assuring the systematic investigation of the root causes of non-conformities, preventing their recurrence (corrective action), and preventing occurrence companywide (preventive action). The FDA frequently cites companies for violation of CAPA procedures. Here are the details of the system from the FDA website [5]:

1. Verify that CAPA system procedure(s) that address the requirements of the quality system regulation have been defined and documented.
2. Determine if appropriate sources of product and quality problems have been identified. Confirm that data from these sources are analyzed to identify existing product and quality problems that may require corrective action.
3. Determine if sources of product and quality information that may show unfavorable trends have been identified. Confirm that data from these sources are analyzed to identify potential product and quality problems that may require preventive action.

4. Challenge the quality data information system. Verify that the data received by the CAPA system are complete, accurate and timely.

5. Verify that appropriate statistical methods are employed (where necessary) to detect recurring quality problems. Determine if results of analyses are compared across different data sources to identify and develop the extent of product and quality problems.

6. Determine if failure investigation procedures are followed. Determine if the degree to which a quality problem or nonconforming product is investigated is commensurate with the significance and risk of the nonconformity. Determine if failure investigations are conducted to determine root cause (where possible). Verify that there is control for preventing distribution of nonconforming product.

7. Determine if appropriate actions have been taken for significant product and quality problems identified from data sources.

8. Determine if corrective and preventive actions were effective and verified or validated prior to implementation. Confirm that corrective and preventive actions do not adversely affect the finished device.

9. Verify that corrective and preventive actions for product and quality problems were implemented and documented.

10. Determine if information regarding nonconforming product and quality problems and corrective and preventive actions has been properly disseminated, including dissemination for management review.

Training Production Operators to Identify Incidents That May Result in Device Defects

One of the best tools for identifying problems before they result in harm is *incident reporting*. This tool requires the reporting of all errors and situations that can result in an adverse event. All hospitals use it; medical device companies should also use it. Operators do make mistakes that can result in an unsafe product and come across situations that can do the same. If they are rewarded for reporting their own mistakes in a blame-free culture, they can report many such situations. Engineers can then design out these errors and situations or find a way to mistake-proof them. If an operator forgets to put a lock washer in an assembly, the employer can buy the assembly from the supplier instead of buying a screw from one supplier and the washer from another supplier, or use a poka-yoke (mistake-proofing) method so that the lock washer will not be forgotten.

Similarly, if an operator finds an instrument that is not properly calibrated, he/she should report it in the incident report in addition to getting the problem corrected.

Production Validation Testing

This is a very powerful tool for identifying hidden production problems. Figure 7.1 shows the results of a wire bonding test used to prove the strength of the bond. When we use a special Weibull graph plot (as in Figure 7.1), it shows two slopes for the strength data; usually the points on the first slope represent production defects that were not caught prior to testing. The second slope represents design defects. This knowledge is a great asset in the hands of engineers. Details of this tool can be found in reference [6].

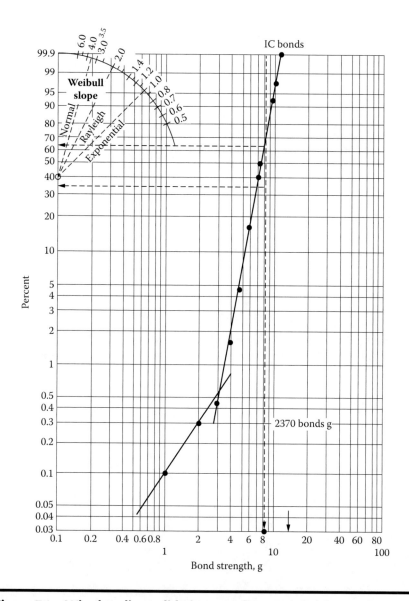

Figure 7.1 **Wire bonding validation test shows production problems (the first slope). (From D. Raheja and M. Allocco,** *Assurance Technologies Principles and Practices.* **New York: Wiley, 2006.)**

Summary

Production defects are responsible for roughly 30% of recalls. Defect prevention and validation testing are very effective in preventing defects from going to customers. We should try to change product design or change the process to eliminate defects permanently. If not possible, we may have to rely on production validation testing. Using such approaches results in a very high return on investment.

References

1. Medtronic, Inc., "FDA Classifies Medtronic's Worldwide Voluntary Field Action on Guidewires as Class I Recall," Press Release, November 15, 2013, http://www.fda.gov/Safety/Recalls/ucm375363.htm?source=govdelivery&utm_medium=email&utm_source=govdelivery
2. Invacare, "Medical Device Field Removal Joystick Recall," October 2013, http://www.invacare.com/cgi-bin/imhqprd/inv_productalert/prod_alert.jsp
3. Bridgestone Tire. *Industry View: How Can Dr. Deming Help You Grow?* http://www.bridgestonetrucktires.com/publications/98v3issue2/v3i2IndV.asp, accessed on March 26, 2014.
4. U.S. Food and Drug Administration (FDA), *Guidance for Industry: Q9 Quality Risk Management.* Rockville, MD: Food and Drug Administration, June 2006.
5. U.S. Food and Drug Administration (FDA), "Corrective and Preventive Actions (CAPA)," June 28, 2010, http://www.fda.gov/ICECI/Inspections/InspectionGuides/ucm170612.htm
6. Raheja, Dev, and Allocco Michael. *Assurance Technologies Principles and Practices.* New York: Wiley, 2006.

Chapter 8

Preventing Software Design Recalls

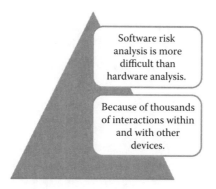

Software risk analysis is more difficult than hardware analysis.

Because of thousands of interactions within and with other devices.

Introduction

Software failures were responsible for 24% of all the medical device recalls in 2011, according to data from the U.S. Food and Drug Administration (FDA), which said it is gearing up its labs to spend more time analyzing the quality and security of software-based medical instruments and equipment. The FDA's Office of Science and Engineering Laboratories (OSEL) released the data in its 2011 Annual Report on June 15, amid reports of a compromise of a website used to distribute software updates for hospital respirators. According to the agency: "The absence of solid architecture and 'principled engineering practices' in software development affects a wide range of medical devices, with potentially life-threatening consequences ...". Recent research done on the security of medical devices by a team of researchers identified software security vulnerabilities in software that controlled an auto-mated external defibrillator (AED), which is used to treat cardiac arrhythmias. The researchers also found that the device would accept unsigned, counterfeit software updates [1].

Software Requirements Analysis

Software intelligence in most devices is a function of several systems working together to produce properties and behavior different than those of the components. The discipline of gathering such intelligence is often missing. This is one reason for the increase in device recalls. Most manufacturers have not applied rigorous hardware risk analysis to software designs. The same methods apply to software even though there are differences between software and hardware. Software requirements analysis, Preliminary Hazard Analysis (PHA), Failure Mode and Effects Analysis (FMEA), Fault Tree Analysis (FTA), and Hazard and Operability (HAZOP) are great tools for software safety.

Approximately 80% of the dollars that go into software development are spent on finding and fixing failures. This is very inefficient and costly. For a robust design, the opposite is required; that is, 80% of dollars should be spent on preventing failures so that the chance of device recalls is dramatically reduced. The need for a cost reduction is obvious; a fault that is fixed during the concept stage costs only a small fraction compared to the warranty costs and the cost of downtime for fixing it later. The requirements analysis allows the product development teams to see the forest instead of the trees by capturing the complexity of software requirements. Lack of this macrovision is bound to result in many costly failures.

Requirements analysis captures three types of risks: the risk of the *known*, the risk of *known unknowns*, and the risk of *unknown unknowns*. The known risks are understood through past history and customer needs statements. Unfortunately, customers are not aware of more than 60% of the potential requirements until a device is functionally broken. This reflects the author's experience working with many organizations. They recognize the need for a requirement only after they "don't get it." This is particularly true for software specifications. Many companies attempt to make use of lessons learned, but most do not have formal and verifiable protocols. Some known risks can be identified through tools such as FMEA, FTA, and Event Tree Analysis (a tool used in the nuclear industry). Therefore, some progress is being made in handling the known risks. The other two risks are significant, but there has been no significant progress in handling them.

The known-unknown risks are unknown to the specification writer, but are known to users of similar devices. The author, while working with the Baltimore Mass Transit System, could not come up with more than 200 requirements in a specification with the engineers. He interviewed train drivers, technicians, and passengers in San Francisco's BART system, and discovered a list of more than 1,000 concerns. At least 500 of them were added to the Baltimore requirements.

The unknown unknowns are special risks. They mostly apply to smart devices such a smart infusion pumps, MRIs, patient monitoring systems, and smart alarms that depend on trustworthy interoperability. The faults are usually unpredictable with the tools we have today. The reason is that the systems are too complex. No longer are we dealing with one mechanical system that can perform and stand alone. The software in a pacemaker may require more than 80,000 lines of code, a drug-infusion pump 170,000 lines, and an MRI scanner more than 7 million lines. This growing reliance on software causes problems that are familiar to anyone who has ever used a computer: bugs, crashes, and vulnerability to digital attacks [2]. The key point is that we are dealing with a system made up of several systems. The software typically interacts with several systems, resulting in hundreds of possible interactions, and is called a *system of systems*. The interactions are unbounded. We cannot know how the system of systems will behave by knowing only the behavior of its individual systems. Tweaking one system without knowledge of intersystem behavior is doomed to failure. The unknown-unknown risks are the result of a lack of knowledge of the interactions and associated behavior of the system of systems. Altering the behavior of any part affects other parts and connecting systems. A cross-functional team from interconnecting systems must be involved in thinking of the failure modes of the interactions. Unknown-unknown risks arise from unanalyzed interactions.

In order for software requirements to be comprehensive, the system requirements must be comprehensive. The software requirements in turn require an understanding of complementary properties, capabilities, and behavior resulting from interactions of parts and other systems. In principle, all complex systems are comprised of subsystems that are themselves systems. Every system, therefore, is a system of systems. The categories in the system specifications should include the following:

- Application environment
- Functions
- Active safety
- Reliability
- Safe defaults for sudden malfunctions
- Serviceability/maintainability
- Human interface requirements
- Logistics requirements (software changes, maintenance)

- Input interface requirements
- Output interface requirements
- Installation requirements
- Engineering change requirements

Software FMEA

A software specification is a compilation of all the functions. Each function is analyzed for its failure modes. A failure mode is the answer to the question: "What can go wrong in carrying out this function?" What happens if the function fails? Some failure modes will be obvious, such as "the software may command unintended breathing," or "the function may not execute at the right time."

Omission of the software FMEA is a major source of serious failures and product recalls. The methodology is not yet standardized. Experience shows that almost all companies find it very difficult to directly apply existing FMEA procedures, which were primarily written for hardware. At the functional level, it should make very little difference whether the analysis is for hardware or for software. The failure definition for both is identical. A software failure is the inability of the software to perform its required functions, same as the hardware, but the popular view is that software is different. The reason for this view may be that there is no industry agreement on the causes of software failures and the mission of the analysis. Some analyses include all the causes of software failure, such as hardware anomaly or electromagnetic interference. Others only accept errors in the code as a cause of software failure. Then there are others that include software failures from sensor malfunctions, hardware faults, and software security.

Depending on the application severity, the tool can be applied at the system functional level, the logic level (detailed design), the module level, or at the code level. For safety-critical software, the tool should be applied at all levels. The analysis can also be very useful in developing diagnostics and prognostics requirements.

Analysis at the logic level can be done when the detailed concept design is available (prior to coding). It can be done for robustness of interfaces (interface FMEA), for fault isolation by service technicians (serviceability FMEA), and for software maintainability, safety, and recoverability. One may choose some or all depending on how robust the performance must be.

The functional-level FMEA is the most powerful and cheapest. It is the most powerful because it thoroughly analyzes how all functions can be delivered reliably and safely. It is the cheapest because, at this early stage, the cost of many changes is only a few strokes on the computer keyboard.

Software Interoperability Analysis

The next step is to develop an interactions matrix consisting of all the system requirements (these are technically the functions of the system) in rows and different interacting systems in columns as shown in Figure 8.1 for an automotive transmission [3]. Such a matrix helps in understanding all the interactions during design reviews and helps in identifying the effect of software engineering changes on different parts of the system. In this matrix, "X" indicates a relationship between a function and a subsystem; "C" represents a safety-critical relationship.

Testability Analysis

Software testability can be defined as designing for ease of testing. Each company should define robustness before designing for testing. It may be defined as the ability of software to diagnose faults no matter what caused the software to malfunction and includes the ability to isolate the faults. The

Function	Driver	AMT	Power Train	Hydraulics	Clutch System	X/Y Shifter	Trans. Harness	ECU	CAN Bus	OEM Interface
Engage clutch	X	X	X	X	C	X	X	C	X	X
Engage gear	X	X	X	X	X	X	C	C	X	X
Isolate fault	X	X	X	X	X	X	X	X	X	X
Reduce no fault founds	X	X	X	X	X	X	X	X	X	X
Smooth shifting	X	X	X	X	X	X	X	C	C	X
Starting the engine	X	X				X	X	X		
Acceleration	C	X	X	C	X	X	X	X	X	C
No unintended motion		C	X	C	X	X	C	X	C	X
Robust OEM communication	X	X		X	X	X	X	X	X	X
Reliability centered maintenance	X		X	X	X					
Speed control on grade	C	X	C	X	X	C	X	C	X	X
PTO operation	X	X	X	X		X		C	C	
Maneuverability	X	C	X	X		X	C	X	X	
Low temperature operation		C	X	X			X	C	C	X
Speed control (cruise)	X	C	X		X	X		C	X	
Diagnostics		X	X	X	X	X	X	X	X	X

Figure 8.1 Example of an interface matrix showing the relationships among functions and subsystems. (From Paul Roberts, "FDA: Software Failures Responsible for 24% of All Medical Device Recalls," Threatpost, June 20, 2012, http://threatpost.com/fda-software-failures-responsible-24-all-medical-device-recalls-062012.)

test design should also be able to test that a desired state is reached in case of a software fault. If a desired state is not identified in the specification, the software must always go into fail-safe state. The ability to test right things at high confidence should be the requirement. We need testability analysis with questions such as

- Can we test the safety critical functions with high confidence?
- Can we test to identify the source of the failure at module level so we can design to fail in a safe state?
- Do we have the right test cases to prove that the device will always go into a safe state in case of a hardware failure or malfunction?
- Can we test to isolate faults at code level and result in a safe state if there is a failure?
- Can we test for early prognostic warning?
- Can we test for faults coming from interacting devices?
- Can customers use a test to recover quickly from faults?
- Do we have code-step log files to track where a fault or error occurred?

Such information gives the designer tips on limiting the number of lines of code in each module to minimize the number of paths. The software design may need ways to monitor the faults.

Selecting Software Structure and Architecture

A software structure is like the main structure of an office building, consisting of a foundation and main pillars, or like an organization chart. It usually does not change. Software architecture is like the locations of rooms, size of rooms, plumbing system, wiring system, and water system. These are flexible and can be changed from time to time. For safety-related devices, a top-down structure is usually chosen. The top-level system is decomposed to include top-level subsystems. Then each subsystem is decomposed in the same manner, all the way down to modules.

This structure is least susceptible to interaction faults because each vertical branch is like a separate product. These branches interact only in the top-level module. For example, the four subsystems at the top level for an x-ray machine can be the electrical subsystem, mechanical subsystem, x-ray functions subsystem, and user-interface subsystem. Each is an independent entity. None of them communicate with each other directly. They communicate through a supervisory program at the top level. They can all be developed simultaneously with an independent specification for each. They can interact only downward or upward in each vertical branch. That is why this structure is called a top-down structure.

The architecture involves choosing the size of each module so that there are very few paths that interact within it, and so that there is module independence (a change in a module does not affect other modules), fault tolerance (redundancy to ensure two independent ways of arriving at the right answer), or defensive programming (the program knows in advance where the next destination is (if a command does not go to the predetermined destination because of a change, it alerts the user).

For more information on structures and architectures, see references [4] and [5].

Precautions for Off-the-Shelf Software

Off-the-shelf (OTS) software is commonly being considered for incorporation into medical devices as the use of general-purpose computer hardware becomes more prevalent. The use of OTS software in a medical device allows the manufacturer to concentrate on the application software needed to run device-specific functions. However, OTS software that is intended for general-purpose computing may not be appropriate for a given specific use in a medical device. The medical device manufacturer using OTS software generally gives up software life-cycle control, but still bears the responsibility for the continued safe and effective performance of the medical device. The FDA has the following guidance on OTS software [6]:

> Because the risk estimates for hazards related to software cannot easily be estimated based on software failure rates, CDRH [Center for Devices and Radiological Health] has concluded that engineering risk management for medical device software should focus on the severity of the harm that could result from the software failure. Hazard Analysis is defined as the identification of Hazards and their initiating causes [IEC 60601-1-4]. Based on the definition of Risk Analysis in ISO DIS 14971 and EN 1441, hazard analysis is actually a subset of risk analysis; because risk analysis for software cannot be based on probability of occurrence, the actual function of risk analysis for software can then be reduced to a hazard analysis function. Technically speaking, the use of either term risk or hazard analysis is appropriate. However, CDRH has chosen to use the term hazard analysis to reinforce the concept that calculating risk based on software failure rates is generally not justified, and that it is more appropriate to manage software safety risk based on the severity of harm rather than the software failure rates.

The FDA guidance referenced above provides the software decision schematic in Figure 8.2.

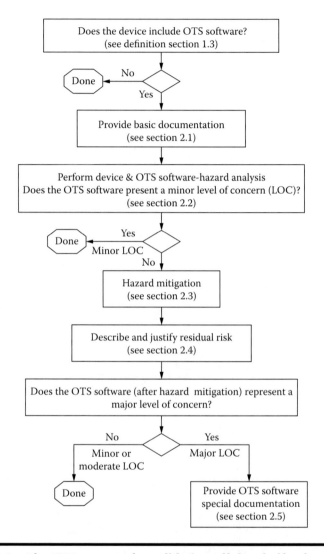

Figure 8.2   The FDA process for validating off-the-shelf software in the devices.

Designing to Minimize User Interface Risks

User-friendly human interface design is especially required for devices performing complex tasks such as surgical robots. Even a well-established premier company, Intuitive Surgical, had a Class II recall. Its famous surgical robot da Vinci was recalled because of its difficult human interface. Results from a small survey of physicians released by the FDA on November 8, 2013, raised questions about training procedures. Among other things, the 11 surgeons responding to the survey suggested da Vinci's complicated interface was difficult to master and that training methods were inconsistent [7].

Reference [8] provides user interface issues and the reasons for new risks. These should be a part of design reviews.

Common User Interface (UI) Issues

- UI complexity causes the user confusion, delay in use, or inability to use the device.
- The UI makes it difficult for the user to correct data entry errors or modify device settings in a timely fashion.
- The UI falsely causes the user to believe a critical situation exists when it does not, or vice versa.
- The UI does not draw attention to dangerous conditions in device operation or patient status.
- The UI does not prevent known likely data input errors.

Common Reasons for Use Errors

- The use environment has negative effects.
- The demands associated with use of the device exceed the user's capabilities.
- Aspects of device use are inconsistent with the user's expectations or intuition.
- The device is used in unexpected ways.
- The device is used in inappropriate but foreseeable ways, for which adequate controls were not applied.

Summary

Software design methodology is most similar to hardware design, even though there are differences. The tools PHA,

FMEA, and FTA have not been applied by many device companies. They should be.

References

1. Roberts, Paul. "FDA: Software Failures Responsible for 24% of All Medical Device Recalls," Threatpost, June 20, 2012, http://threatpost.com/fda-software-failures-responsible-24-all-medical-device-recalls-062012
2. *The Economist*, "When Code Can Kill or Cure," June 2, 2012, http://www.economist.com/node/21556098
3. Raheja, Dev. "Writing Software Specifications through a System-of-Systems Matrix." Paper presented at the SAE Annual Conference, Detroit, MI, April, 2007.
4. Bishop, P.G. *Dependability of Critical Computer Systems 3*. New York: Elsevier Applied Science, 1990.
5. Ganesan, D. et al. "Architecture Reconstruction and Analysis of Medical Device Software." Paper presented at the 9th Working IEEE/IFIP Conference on Software Architecture, Philadelphia, PA, June, 2011.
6. U.S. Food and Drug Administration (FDA). *Guidance for Industry, FDA Reviewers and Compliance on Off-the-Shelf Software Use in Medical Devices*, September 9, 1999, http://www.fda.gov/MedicalDevices/DeviceRegulationandGuidance/GuidanceDocuments/ucm073778.htm
7. PR Newswire, "As Da Vinci Robot Lawsuits Mount, Bernstein Liebhard LLP Comments on Latest Da Vinci Robot Instrument Recall," December 4, 2013, http://www.prnewswire.com/news-releases/as-da-vinci-robot-lawsuits-mount-bernstein-liebhard-llp-comments-on-latest-da-vinci-robot-instrument-recall-234492491.html
8. Story, Mollie Follette. "The FDA Perspective on Human Factors in Medical Device Software Development." Presentation at IQPC Software Design for Medical Devices Europe, February 1, 2012, http://www.fda.gov/downloads/MedicalDevices/DeviceRegulationandGuidance/HumanFactors/UCM290561.pdf

Chapter 9

Preventing Supply Chain Quality Defects to Avoid Recalls

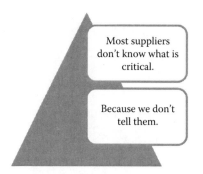

Introduction

Figure 1.1 in Chapter 1 showed that component specifications and their implementation are responsible for about 30% of device failures. We rarely tell suppliers all they need to know. We often leave out reliability, safety, and durability requirements. Some companies don't tell suppliers about critical design features and what is critical to quality. What we learned from previous chapters is, if we don't specify what we want, then we can get something we *don't* want! This chapter will cover these critical needs.

Writing Good Supplier Specifications

An atomizer device sprays liquid medication in aerosol form into the mouth to help a person to breathe. Due to a manufacturing defect in the atomizer, a washer could become dislodged, causing the user to swallow or choke on it. The U.S. Food and Drug Administration (FDA) issued a Class I recall [1].

Let us assume that the atomizer assembly is a purchased part. We don't know if reliability and durability were included in the specifications. Inclusion of either one may have prevented the recall. Another thing that might be missing is the identification of the items/issues that are critical to quality (CTQ). One of them is obviously that the washer should never loosen. These are very common mistakes in supplier specifications.

So what is the solution? The following architecture for writing supplier specifications will help.

Step 1: Develop a checklist covering the items that should be in the specification, such as
  – Functions of the component
  – Reliability
  – Durability (minimum life)
  – Safety-related critical design features (example: no part should break or become loose)
  – Chemical properties of the material (if critical for performance)
  – Critical dimensions
Step 2: Either the customer or the supplier should perform a process FMEA. Doing so jointly is always more valuable. The objective is to make sure that all the requirements in

the checklist will be met consistently over time and that variation in the process will meet Six Sigma quality levels.

Step 3: Conduct an accelerated reliability and durability test on safety-critical design and manufacturing features (where applicable).

Step 4: Mitigate the risks discovered by making design or process changes.

Step 5: Conduct periodic ongoing reliability tests (ORTs) and durability tests, because processes change over time. So do the production operators. Revise the specification if the risks are still high.

The Art of Identifying the Features Critical to Quality

Designers cannot identify all the features that are critical to quality (CTQs). They must take advantage of a cross-functional team that includes the supplier and the customer. The supplier assembly or a component may contain software. In that case, we need to identify the CTQs for software also.

The architecture discussed above also applies to purchasing production equipment. This author was a consultant on a robot purchased from a supplier. The all-day Failure Mode and Effects Analysis (FMEA) with the supplier resulted in over 50 design changes! The cost of the robot increased by about $100,000 dollars, but the potential savings in quality costs was at least 10-fold in 5 years. The problem with robots for automated production is that without proper specifications, they can produce defects faster than humans!

The best way to identify missed critical design features is to conduct accelerated life tests. This will not only identify production defects but also long-term design defects, which are a major factor in recalls.

Assessing Variation in Supplier Quality

Before we choose a supplier, we should have knowledge of their process capabilities. If the process output is not capable of staying within the middle 50% of the engineering tolerance, then we don't have the right supplier for the safety-critical component. Sometimes the upper end of the tolerance is more desirable, such as the strength of a pacemaker wire. In that case, the process must be capable of staying within the upper 50% of the tolerance band. If the low end of the tolerance is desirable, such as with contamination, then the process must stay in the lower 50% of the band.

Once this is done, we can ask the supplier to periodically send the measurements on a sample of about 30 pieces, plotted as a histogram. Compare the histogram width (process capability) with the engineering tolerance. The histogram width should be within 50% of the tolerance band. If the process is capable, you may not need to inspect items on receipt at all for defects that are not safety critical. A good process will have a very low probability of finding a defect even if you do inspect when pieces are received.

Supply Chain Control by Suppliers

Just as we have suppliers, our suppliers *also* have suppliers. It is easy for *us* to overlook the performance of *their* suppliers. They may change processes or change material composition without risk analysis and not know the CTQs. Our primary supplier has the responsibility to ensure the performance of second-tier suppliers. We should include this assurance in quality audits of suppliers.

Assuring Reliability in Performance

The real test of a purchased component is to assess its long-term reliability when it is assembled into the final product. Test the component for reliability as an integral part in the assembly. If the component does not meet reliability requirements, then expect warranty failures or postwarranty failures. Either way, we may lose some customers. Remember that the purchasing department should not focus on the lowest price; they should focus on highest value at a reasonable price for safety-critical parts. The value is the sum of the component price, warranty costs, and the cost of loss of goodwill.

Summary

Most suppliers don't know what is needed, because we don't tell them. Make sure to tell suppliers about reliability, durability, and CTQs. For safety-critical defects, we should try to eliminate the defects by design or by choosing a better process.

Reference

1. U.S. Food and Drug Administration (FDA). "Nephron Pharmaceuticals Corporation EZ Breathe Atomizer," Class I recall, April 2013, http://www.fda.gov/MedicalDevices/Safety/ListofRecalls/ucm355016.htm

Chapter 10

Preventing Recalls Using a Verification Process

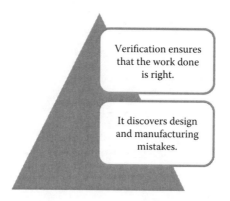

Verification ensures that the work done is right.

It discovers design and manufacturing mistakes.

Introduction

The purpose of verification and validation is to hunt for oversights and omissions in design, manufacturing, and servicing. Who is verifying and who is validating are very important considerations. An independent, unbiased participant with full understanding of risk analysis would be ideal. Finding this person is not always viable. The next best thing is to have a small team (5 to 7 people) get refresher training in risk management, and have well-defined and agreed-upon verification and validation goals.

Independent Verification during Specification Approval

There is a difference between verification and validation. Verification is about confirming that high-quality work was done right, at the right time, by qualified verifiers. If those who write the specifications are not trained in writing specifications based on risk assessment, they are likely to leave out many requirements. One safeguard against too many oversights and omissions is to develop an internal checklist and let the independent reviewer add more to it from his/her personal experience. Here are some questions that should be in the checklist [1]:

- Were the inputs from users addressed?
- Was a search made on adverse events on similar devices on the Manufacturer and User Facility Device Experience (MAUDE) database?
- Is the right intent evident in the Preliminary Hazard Analysis (PHA), Failure Mode and Effects Analysis (FMEA), and Fault Tree Analysis (FTA) documents?
- Are the results of the PHA implemented in the specifications?
- Are the results of the functional FMEA implemented in the specifications?
- Are the software risks analyzed and mitigated?
- Is reliability specified including time duration and worst-case environment?
- Is durability specified in terms of minimum life?
- Are fail-safe requirements addressed?
- Is the device designed to isolate faults and inform users about specific problems?
- Has redundancy been considered to avoid critical failures and malfunctions?
- Is reliability-centered maintenance specified for complex systems (MRI, nuclear medicine)?

- Are sterility issues in packaging addressed?
- Does the device need to withstand earthquakes, loss of power, and unusual loads?
- Is the device tested for reliability in overseas environments?

Independent Verification during Final Design Approval

Many companies have a so-called *final design review.* Some call it the *critical design review.* This is the last chance to determine if anyone is concerned about anything in the design. Always remember: *Learn to say "no" to yes-men.* If the final design is presented and everyone votes "yes" and approves the design, then *your* answer should be "no." Why? Because there are almost always new problems lingering in the minds of the team members, but they don't speak because they think it is too late to interfere. If no one is challenging the design, the device is bound to crash in use, unless the device is very simple. No matter how good the design is, an independent facilitator can find many issues with it. Ford Motor Company hired a new vice president during the design of the 1995 model of the Lincoln Continental. The company had been making this car for years and everyone on the team had at least 10 years of experience. The design was already approved, but the new vice president insisted on questioning every detail of the design with a cross-functional team made up of engineers from each subsystem. He included a marketing manager and service engineers. They made over 700 changes in the design! Since they made these improvements before the design was in manufacturing, they saved $60 million (the potential cost of making engineering changes in production, rework, and reducing the number of inspectors). The new product was released 4 months ahead of schedule!

Independent Verification during Pilot Production Approval

It is very important to do a good job when planning the verifications and validations that are needed before approving the production process. The verification phase consists of going through the design FMEA and the process FMEA to make sure the problems are designed out as much as possible, and dependency on inspection and testing is minimized. Verification includes evidence that the process is capable and consistent over time without producing defects. If defects are produced, there must be a system to detect them. In unusual cases, 100% inspection may be required, but that much inspection should *not* be the norm. Good verifiers analyze operator mistakes that can be made and may not be caught. They understand that inspectors will not be able to discover *all* these mistakes. Some inspections can be automated, which is better than humans judging the quality. Some mistakes operators often make are

- Errors of substitution: the operator may pick a wrong fastener unknowingly.
- Errors of selection: the operator may press the wrong button for controlling a chemical composition.
- Errors in reading: operators see/read what they expect rather than what it says. Operators may erroneously see a 10–100 scale on an instrument, assuming that it is a 1–10 scale.
- Errors of irritation: operators may assemble parts poorly because of too many distractions.
- Errors of warning: the operator does not remember all the cautions and warnings in the equipment instruction manuals.
- Errors of alertness: operators make mistakes because they are tired or have family problems.

- Errors of lack of understanding: an operator may not understand the process and may be afraid to ask questions.
- Errors of haste: the operator cannot finish assembling in the allocated time, and takes shortcuts such as not tightening the fasteners properly.
- Errors of reversal: the operator may attach a component backward or upside down.
- Errors of sequencing: the operator may not follow the approved sequence in doing work.

A typical problem is that many FMEAs recommend inspection or production testing in the Action column instead of changing the design or the process. This action (or inaction) opens possibilities of recall because production operators are human and are sure to make mistakes. Tests in production often do not duplicate the use environment and cannot replace inspection. Therefore, there will be device failures during validation tests and engineers will be unable to establish confidence that defects will not be passed on to customers.

Independent Verification of Supplier Quality Assurance

Supplier defects can be a major source of product recalls. Independent verification of the supplier's product to find defects at the time that the product is received is rare. If defects are found frequently, it is a symptom that the supplier's processes are not capable of producing a product that is within the 50% tolerance range. It is better and cheaper to pay more for a qualified supplier than to pay for questionable components for years.

The enemy is not actually the supplier. As we discussed in an earlier chapter, we are our own worst enemies. We don't tell suppliers all they need to know. Verifiers should ensure that suppliers are told about reliability, durability, and the things that are critical to quality (CTQs). If a supplier has a capable process, then the next verification should be whether we can rely on the supplier without inspecting its product when we receive it. In many companies in Japan, parts go straight to the production floor. If the supplier has a good process that is capable of defect-free production, the receiving inspection should be unnecessary. Give inspection money to the supplier to design a more robust process. This approach was covered earlier with the paradigm, *no inspection needed is the best inspection*!

Verifying Day-to-Day Control in Production

One strategy we must follow, regardless of how much it costs, is verifying how well our daily controls conform to U.S. Food and Drug Administration (FDA) regulations. If we don't have the basic discipline to following good design and manufacturing practices, the rest of the effort is not going to work. All we have to do is to look at the FDA violations. They can be accessed on the FDA website by searching for warning letters.* Here is a checklist developed from these warnings:

- Are complaint-handling procedures well implemented?
- Are procedures for implementing corrective and preventive actions well established and maintained?
- Are the corrective and preventive actions validated with a high degree of assurance that the risk is prevented?
- Are procedures for acceptance of incoming product well established and maintained?
- Are procedures for a process that cannot be fully verified by subsequent inspection and testing (as required by 21

---

* This section on the FDA website covers the requirements for process evaluation.

CFR 820.75(a)) adequate for providing a high degree of quality assurance?

■ Are procedures for submitting Medical Device Reporting (MDR) documents within 30 days in the case of an adverse event well implemented?

■ Are there procedures to adequately maintain an MDR event file as part of your complaint file, as required by 21 CFR 803.18(e)?

■ Are there written procedures for validating device design as required in 21 CFR 820.30(g)?

■ Do we have sufficient personnel with the necessary education, background, training, and experience to ensure that all activities required by 21 CFR 820 are correctly performed, as required by 21 CFR 820.25(a)?

■ Are procedures to ensure that all purchased materials conform to specifications as required by 21 CFR 820.50 written and updated?

■ Are procedures for quality audits well implemented as required by 21 CFR 820.22?

■ Are design control procedures well documented and implemented as required by 21 CFR 820.30?

■ Are there written procedures for changes to a specification, method, process, or procedure and for verifying or validating that change before implementation as required by 21 CFR 820.70(b)? How well are the corrective actions documented and implemented?

■ Are there written procedures for sampling plans based on a valid statistical rationale and to ensure that sampling methods are adequate for their intended use, as required by 21 CFR 820.250(b)? How well they are implemented?

■ Are there adequate procedures for analyzing processes, work operations, concessions, quality audit reports, quality records, service records, complaints, returned product, and other sources of quality data to identify existing and potential causes of nonconforming product, or other quality problems, as required by 21 CFR 820.100(a)(1)?

- Are there adequate procedures for finished device acceptance to ensure that each production run, lot, or batch of finished devices meets acceptance criteria, as required by 21 CFR 820.80(d)?
- Are there adequate procedures for receiving, reviewing, and evaluating complaints by a formally designated unit, as required by 21 CFR 820.198(a)?
- Is there an adequate procedure to submit a written report (within 10 days) to the FDA of any correction or removal of a device initiated by a manufacturer as required by 21 CFR 806.10(b)?
- Are there adequate procedures to ensure that only those devices that are approved for release are distributed?
- Are there adequate procedures to fully document your validation activities for *in vitro* samples such as biopsies?
- Are there adequate procedures to process your complaints in a uniform and timely manner?

Summary

Verification ensures that the work done is right and at the right time. It discovers design and manufacturing mistakes and should include verifying suppliers' inherent ability to deliver defect-free components and assemblies. The most productive verification is to look up the FDA warnings to manufacturers and make sure that your company is not violating any of those warnings.

Reference

1. Raheja, Dev, and Allocco, Michael. *Assurance Technologies Principles and Practices*, New York: Wiley, 2006.

Chapter 11

Preventing Recalls Using Design Validation Process

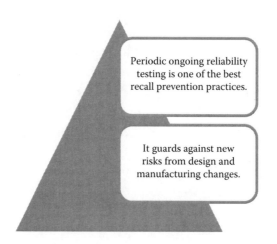

Periodic ongoing reliability testing is one of the best recall prevention practices.

It guards against new risks from design and manufacturing changes.

Introduction

Validation is about requalification of the design produced by real production processes, with real operators, and with real users! The purpose of validation testing is much more than making sure that the device conforms to the specifications. It can be used for finding design and manufacturing mistakes that can result in recalls. Therefore, the test has to be planned accordingly. It may seem like too much testing, but it is needed to find hidden problems. The cost of overtesting on a sample can save millions of dollars later.

Design Validation Testing for Reliability

To understand reliability, it is necessary to understand the failure rate profile for the life of the device; this is called the *bathtub curve* and is shown in Figure 11.1. For electrical and electronic devices, it usually consists of three regions, shown in profile (a). Each region has a different failure rate profile. Region I usually represents the first two to three years when defects from manufacturing or gross oversights in design show up. Because this region is from manufacturing or design flaws, it is popularly called the infant mortality region. Then the device has a constant failure rate resulting from unusual random loads such as lightning and thunderstorms. This is represented by region II. Finally, the components wear and the failure rate increases rapidly, as shown in region III. An easy-to-understand example of a random failure would be a tire on a car failing from a nail cutting into it; it is random and an

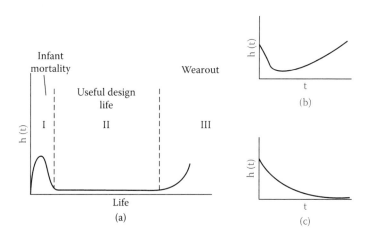

Figure 11.1   Failure rate profiles for electrical products, mechanical products, and software products.

unusual load. For mechanical devices, failure profile (b) is common. For software, profile (c) is the case.

The reliability test occurs in region I, and the test failure rate decreases with time until it reaches a constant or very low failure rate. To prevent recalls, we must make sure safety-related failures are zero. As mentioned, this test continues until the failure rate is very low. The ideal procedure is to plot the data over time (such as documenting the failure rate at 100-hour intervals), and continue until the failure rate is very low. The failure rate equation is

Failure Rate for a chosen interval = Number of Failures/ (Number of test devices remaining at the beginning of each test interval)

Example: Let us assume that the data is plotted in 100-hour intervals with a starting sample size of 50 devices. Assume that 12 devices fail in 300 hours. If 2 devices fail in the next interval (300–400 hours), the failure rate is

Failure Rate = $2/(50 - 12) = .0526/100$ hours or = .000526 failures/hour.

For purely mechanical devices, profile (b) in Figure 11.1 shows that there are only two regions. The first region, the decreasing failure-rate region, is from manufacturing and supplier defects. It may take a few years to bottom out. There is usually no constant failure-rate region, although it is possible for designs with a long life. The product then starts to wear as a result of worn components. The usual strategy is to design the product for much higher loads and to reduce the test time substantially by using test stresses that produce a high acceleration factor. This author designed a product to withstand four times the load (Chapter 5). The test time was reduced to hours instead of weeks. If such a test is done long enough to trap manufacturing faults (first region in the bathtub curve), it is called Highly Accelerated Stress Screening, or HASS.

Design Validation Testing for Durability

It is worth repeating that design validation for durability is requalifying for durability using production parts and operators. The test can be for an equivalent of 20 years of life for an MRI, 5 years for infusion pumps, 10 or more years for pacemakers, and some customers will use such products longer. The law expects a device to be safe for as long as the consumer (patient) of the device uses it. We cannot test for that long a duration at normal loads, so we need a strategy to predict life. The life-prediction methods were discussed in Chapter 5. For more details, see reference [1].

A design validation test should be repeated at a set frequency, such as once in six months or once a year, depending on the complexity of the device and the risk of a recall. A product was recalled because a gasket seal in the device's relay leaked. Upon investigation, it was found that the device maker used aerospace standards for durability. That is the best test anyone can use. The problem was that the durability validation was done only once at the time the new device was introduced to the market. After about 4 years, the production tooling wore out and the gasket surfaces on the relay were not smooth anymore. That is why the seal leaked. If a durability validation had been done at least once a year, such defects would have been discovered before the risk became high. For this reason, a periodic validation for durability is essential. A great lesson to be learned.

Design Validation for Safety

The prerequisites for testing for safety are validation tests for reliability and durability. Usually, each device failure is a potential hazard because it can interrupt the care of the patient. There are at least five more tests for device safety as applicable:

1. The device is capable of going to a safe state when it fails or malfunctions.
2. The device protects the patient from harm by giving warnings or alarms in case of a human error, including maintenance errors.
3. The device can be used safely by inexperienced users.
4. The device protects the patient if an overconfident user, such as the doctor or a nurse, overrides alerts when in a rush or when distracted.
5. The device is safe to use in different environments, such an alarm should be loud enough in a noisy intensive care unit.

The third item—the device can be used by inexperienced users—requires a special test called *usability testing*. It should also include experienced users, because they can be overconfident and take shortcuts. They often ignore warnings. This is especially true when a device is modified and the experienced user does not see the need for reading instructions or may not know details about the modification. They may not be familiar with the details in instruction manuals. Users may include doctors, nurses, support staff, patients, and families of patients.

It is wise to perform usability analysis first and use this document to develop usability testing. Usability analysis was covered in the Chapter 2. The following issues were found.

- Battery runs out.
- Pump stops working.
- Patient circuit unhooked.
- Circuit blockage.
- Pressure delivered is too high or too low.
- Pressure sensor failure.
- User interface failure.
- Loss of ground affecting other equipment.
- Prolonged exposure to high temperature in desert environment.
- Malfunction in prolonged freezing temperatures.

- Unusual power surges.
- Water intrusion resulting in electrical malfunction.
- Someone turns off the alarms.
- Someone turns off the LEDs.
- Device stored in extreme environments.
- Device is carelessly dropped.
- Device is thrown with extra force during transportation and shipping.
- Placement of heavy materials on top of the device during transportation.
- Fine sand intrusion in a desert environment.

Let us see how we can use this list to design the test. Going through the issues in this list, insert faults into the device and prove that nothing bad will happen to users. For example, tests may consist of the following actions:

- Suddenly remove the battery while the device is in use.
- Disconnect the pump while the device is in use.
- Disconnect the electrical circuit while the device is in use.
- Block the circuit while the device is in use.
- Deliver high pressure while the device is in use.
- Deliver low pressure while the device is in use.
- Disconnect the pressure sensor while the device is in use.
- Fail the user interface screen while the device is in use.
- Disconnect the ground wire while the device is in use.

We need to develop such test cases for all of the issues. If nothing bad or unexpected happens, we have a good safety qualification. If something negative happens or we discover new risks, then the specification and design must be corrected.

Using Field Validation to Identify New Risks

The tests done in design and in production cannot be fully trusted because they often do not create real and unusual use scenarios. Therefore, the validation is not complete until there is documented evidence that there are no new risks in field use. This is called the *final validation*. It is very critical because we must understand the unidentified risks.

As we discussed, new defects can creep in between the consecutive tests of durability and reliability. Users often find new production defects and sometimes new design defects that engineers overlooked. If we can get this information as early as possible, we can prevent problems quickly. One company's salesmen send customer complaints immediately using optical scanners. This information goes to data analysis software, which generates an automatic engineering change notice as soon as it detects two similar complaints. Why generate an engineering change for production defects? The reason is that the design engineers can eliminate many production defects by changing the designs. They can eliminate fasteners in the design if fasteners come loose. They can eliminate welding if welding is a problem, and so on. If they cannot eliminate the risk, they should generate a process change order.

Summary

Periodic testing for reliability and durability is one of the best practices for avoiding recalls. The tests must be well thought out based on usability analysis.

Reference

1. Raheja, Dev, and Allocco, Michael. *Assurance Technologies Principles and Practices*. New York: Wiley, 2006.

Chapter 12

Recall Planning to Maximize Efficiency in the Event of a Recall

Good planning is about the solution first, no problem later!

If the plan includes right things early, the results will be cheaper, better, and faster!

Introduction

According to the ExpertRECALL Index, a report that aggregates and tracks cumulative recall data from the Consumer Product Safety Commission and the U.S. Food and Drug Administration (FDA), medical device recalls documented in the FDA's first-quarter enforcement reports in 2012 affected nearly 82 million units, a 5-quarter high [1]. Products impacted by the increase included alcohol prep pads, catheters, needles, latex gloves, and other medical device products.

Manufacturers are expected to voluntarily initiate a recall when a device poses a risk to the public. The FDA can initiate or order a recall if necessary, under a certain authority. The FDA provides general oversight of a medical device recall once it is initiated. The recall procedure is covered in the FDA document titled *Guidance for Industry: Product Recalls, Including Removals and Corrections*, which is available at the FDA website (http://www.fda.gov/safety/recalls/industryguidance/ucm129259.htm). The team leader should be very familiar with the process. However, within the company, more preparation is needed. How well the manufacturer prepares in advance is the key to handling the recall without wasting time and resources.

Overview of the Plan

A *recall* is the correction or removal of a device where the FDA finds that there is a reasonable probability that the device will cause serious, adverse health consequences or death. Recall classification is by the numerical designations I, II, and III, which are assigned by the FDA to a particular product recall to indicate the relative degree of health hazard presented by the product. A Class I recall is a situation in which there is a reasonable probability that the use or exposure to a device will cause serious adverse health consequences or death. A Class II recall is a situation in which use or exposure to a device may cause temporary or medically reversible adverse health consequences, or where the probability of serious adverse health consequences is remote. A Class III recall is a situation in which use or exposure to a device is not likely to cause adverse health consequences [2].

To make sure everything right happens at the right time, a written plan should be discussed with a cross-functional team. A plan implies that all the activities required are thought out in advance. The person who will be responsible for the execution of the entire plan on a timely basis is also thought out in advance. Each team member has an assigned responsibility in the plan so that time is not wasted while harm is being done. The plan should include the following elements: immediate recall coordination, review of the discovered risks, review of data management, verification of activities for effectiveness, and proper recall closure.

We can develop a better recall plan by conducting an informal FMEA on recall preparedness. You will be surprised at the number of things that are missing in the preparedness plan. Even better, conduct a drill to verify that everything goes according to the plan, and that the assigned team members know their tasks well.

A major concern of the recall team will be how to deal with a market shortage that will impact users and customers. The plan should include how to prepare for this risk.

Immediate Recall Coordination

The speed of a recall is the key to a successful recall. The first step can be called *damage control*. The recall team has to determine the persons who are in the path of harm, such as patients, hospitals, nursing homes, outpatient clinics, and patients using the device at home. For some devices, caregivers may cause harm by continuing to use the device. If the discrepancy discovered can cause major harm or death, then all those affected must be notified by a safety alert. This is not an easy task. The recall coordinator should have the authority to make timely decisions and muster the resources available. Safety alerts may have to be done with the help of media, such as radio stations, television stations, and newspapers, depending on the severity of the risk. The plan should include which team member is responsible for implementing the safety alert. To ensure the broadest coverage, press releases can include the Associated Press, which can get the alert out in major newspapers.

Lot and serial number traceability at the component level is extremely critical. Traceability should also be tied to engineering changes and manufacturing changes. Many recalls result from both. If we don't have a record of the dates of the changes, we may have to recall all the devices made prior to the change. This information should be readily available. Verify that the team member who is responsible for dealing with the media knows who to call, and knows how to follow up to avoid negative publicity. In case there is already negative publicity, this person should be trained on how to handle it. Ensure that all employees are familiar with the company spokesperson's message. No other employee should speak to the media to avoid a misunderstanding of the alert. Consider who will have responsibility for overseeing incoming and outgoing documents.

Review of the Discovered Risks

Once the potential harm from the device is discovered, risk assessment is necessary to determine the extent of the recall. Depending on the device's degree of hazard and extent of distribution, the recall strategy will specify the level in the distribution chain to which the recall will be extended including wholesalers, retailers, and users and consumers. The reason for a recall is because the risk is higher than the initial assessment or there is a new risk.

Risk assessment requires two inputs: the probability of harm, and the severity of the harm, death being the highest severity. Each company is supposed to have established the definitions of different levels of harm probability and severity according to ISO 14971 (explained in the Chapter 3). The probability of harm is a multiplication of the probability of the hazard in the device and the probability of the hazardous situations. The total impact of the probability and severity of harm is the risk. The extent of the recall that must be implemented depends on this assessment.

If the risk is totally new, a good practice is to develop a fault tree for the causes of the risk. Normally, we look for one root cause, but that is an illusion. There are always multiple causes that can result in the same risk. The same risk can develop again and again until we fix other causes that appear in the fault tree.

Make sure to verify interoperability and integration issues so these issues are mitigated in the corrected device. Many hospitals integrate data from medical devices for remote data display, sending inputs to connecting devices or alarm distribution. The reverse is true also. Your device can be getting data from other devices such as patient monitoring and pain control systems. Assess any integration issues that increase the level of risk. Another issue can be communication malfunctions between the device and the IT systems. An example of this challenge is medical device clock time accuracy—an important aspect of interoperability. What if the clock is a minute, hour, or day off? One may not realize that some electronic medical records (EMRs) are configured to reject data that are outside of a certain time window. The root cause of some problems with missing data in EMRs is the inaccurate time stamp of readings [3].

We may even have to perform a FMEA if it was not done during the design to guard against overlooked interoperability hazards and hazardous situations. Make sure the device is safe to interoperate.

Review of Data Management

A challenging task in managing a recall is determining exactly who has the product, where those persons are, and how to reach them. Users can be traveling or their contact information may be obsolete. The data on users and their contact information is essential and must be current so they can be contacted. The FDA will be overseeing to ensure that the recall management is effective.

Be aware of the complexity of device kits and the hazards relative to the numerous component parts of a medical device, the locations of recalled product, the people who need to be notified, and the audit trails required to verify that steps have been taken. It is clear that automation is the only way to execute a recall quickly and efficiently. Odenkirk and Kozenski have the following conclusions and advice [4]:

> [M]any medical device manufacturers still rely on monolithic enterprise resource planning (ERP) systems that actually slow their ability to track and manage recalls. Some don't even have a warehouse management system (WMS) in place to help them understand where product is in the supply chain. The problem with the ERP approach is that very often those systems require manual processes to track the data manufacturers need to implement a recall effectively—and the more manual the process, the more errors, omissions, and delays can occur. Additionally, legacy systems commonly do not have the ability to track shipment and inventory information at a level granular enough to perform a focused product recall. That's why many medical device and diagnostic manufacturers are examining their existing processes for managing recalls and adapting some of the best practices already proven to work in managing recalls in the automotive, electronics, and retail

industries. Many of these processes rely on business intelligence tools that incorporate analytics, alerting, and reporting with supply chain execution systems that provide visibility across the supply chain. Medical device manufacturers are also going another step further, moving away from managing their own supply chains and using third-party logistics (3PL) providers to manage their supply chains.

Medical device manufacturers have to handle complex domestic and international distribution requirements, as well as compliance with FDA regulations. They ship product to a wildly diverse group of customers—hospitals, integrated health systems, distributors, alternate care locations, sales representatives, and even the federal government— all with varying levels of IT systems and savvy. Tracking all of those customers and how they want to be contacted can be overwhelming if the system isn't set up properly. In a normal business-to-business environment in other industries, a manufacturer can often send out recall notices automatically through their electronic data interchange (EDI) application. However, in the healthcare industry, recall notifications are sent manually because recall notifications via EDI have not yet been adopted across the industry. For instance, sales representatives carrying trunk stock, who spend a lot of time on the road, may not have integrated inventory management systems and would prefer to be notified by a text message. Some physician offices may prefer an email notification when a recall occurs, while others may prefer to be notified by a digitized voice mail. Therefore, it is important to know not only who needs to know, but also how they need to be contacted to get the recall information in a timely manner. The numerous contact points that need to be reached during a recall can add a lot of complexity. Therefore, these notifications should not be managed manually.

Verification of Activities for Effectiveness

The recalling firm is responsible to ensure that the recall is effective. The recall team should verify the effectiveness of all activities. A major part of an effectiveness check is to verify that your recall notification letter was received by those informed and that they understand the contents of the recall letter and are able to follow the instructions. The effectiveness check should also verify that the media are doing a fair job of informing the public.

If the effectiveness checks indicate that the recall notification was not received, then we should take the necessary steps to make the recall effective. These steps may involve sending out a follow-up notification that better identifies the product, better explains the problem, and provides better instructions to customers.

The FDA district recall coordinator provides a copy of the FDA document *Methods for Conducting Recall Effectiveness Checks* (http://www.fda.gov/ohrms/dockets/dockets/76s0006/76s-0006-rec0001-01.pdf). The FDA may also audit the customers of the recalled device to ensure the recall activities are accomplishing their purpose. If the FDA determines that the recall is ineffective, it may ask the manufacturer to take appropriate actions. It may require reissuing recall notifications.

Closing the Recall

The FDA and/or the recall committee determine when the recall is complete. This is when the authorities are satisfied that all reasonable efforts have been made to remove the recalled product and that recalled devices have been destroyed. It is essential that all documentation is complete. There is usually an FDA follow-up inspection of the plant to verify that the reason for the recall has been corrected.

The recall coordinator will also issue a report to the department head and the recall committee as to the reason for the recall and corrective action steps to prevent the problem from happening again.

There are three important activities that need special attention:

1. The physical recovery and disposal of the devices
2. The documentation of the collection and disposal of the devices
3. Risk assessment and acceptability of the solution

Distribution and plant management will need to designate a return pooling area for the recalled product. If this area is within the plant, it must be segregated from normal operations. The recall team must review the scope of distribution for the product and determine the quickest way to get the product out of circulation. All returns need to be clearly marked to avoid any confusion with unrelated products. A disposal site must be found for the recalled product.

Documentation should ensure that everyone involved with the retrieval of recalled product records the numbers and locations from which the devices were retrieved, along with their disposition. This documentation is essential to gain closure and termination of the recall. This documentation must be forwarded to the coordinator.

Summary

The recall process is the combination of a risk management process and project management. The team leader must have both skills. It requires a risk management plan, the ability to expedite, and an understanding of the system of data collection and limitations.

References

1. Michaels, Bob. May 15, 2012, "You're Hit with a Medical Device Recall—Now What?" Qmed, May 15, 2012, http://www.qmed.com/mpmn/medtechpulse/ youre-hit-medical-device-recall%E2%80%94now-what
2. U.S. Food and Drug Administration (FDA). "Recall Activities," in *Investigations Operations Manual 2013*, http://www.fda.gov/ downloads/ICECI/Inspections/IOM/UCM123513.pdf
3. Goldman, Julian M., MD. "Medical Device Interoperability: A 'Wicked Problem' of Our Time, *Patient Safety and Quality Healthcare*, 24–27, January–February 2013.
4. Hartford, Jamie. "How to Handle a Medical Device Recall," Medical Device and Diagnostic Industry (MDDI), May 13, 2011, http://www.mddionline.com/article/ how-handle-medical-device-recall

Chapter 13

Role of Management in Preventing Recalls

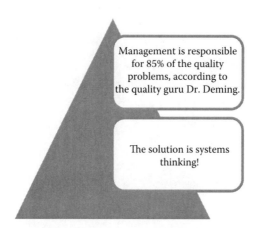

Management is responsible for 85% of the quality problems, according to the quality guru Dr. Deming.

The solution is systems thinking!

Introduction

Why is the number of medical device recalls not going down? According to this author's 30 years of experience as a medical device consultant, a close look at misused and misapplied basic analysis tools shows that the practices are no better than astrology! This is true. Most risk management tools in the ISO 14971 are grossly underused. Take for example the most important tool, the Preliminary Hazard Analysis (PHA).

This tool is rarely used and there is no system representation during brainstorming of system hazards. The system includes suppliers, users (nurses, patients), and the interoperability software designers who connect the device with other hospital functions. According to the world quality guru Dr. Deming, management is responsible for 85% of quality problems such as inadequate policy, inadequate operating procedures, inadequate supervision, inadequate work conditions, lack of team work, and inattention to how quickly problems are prevented.

Management Policies

The first task for management is to set a policy that aims at doing right things, by the right people, at the right time. This policy should be measurable so that progress can be tracked until it is fully working. Some define policy in the form of a mission statement. Most companies have a policy to listen to the voice of the customers and meet all their expectations. This is a very inadequate policy! Why? Because customers don't know all they need to know until the device is broken! We have to do much more than just meeting the customer's explicit requirements.

An ideal policy would be "Nothing bad will happen to the user." The user can be a nurse, a doctor, a patient, or a family member of the patient. Corning Glass Works is an ideal example. Their policy is to always make a *defect-free product*. A production worker discovered that he had not used the right chemical proportions in the production of television glass screens for Sony. He could have kept the secret to himself. There was no way anyone else would have known, not even the customer. Instead, he sent the material to the lab for evaluation. He discovered that the defect would not show up for about eight years, when the screens would start to fade. He reported it to management. Management informed Sony that it would not ship that batch and would reimburse Sony for any losses because of a production stoppage. The surprise in this example is that Sony wanted the shipment despite the defect. Their reasoning was that most customers buy a new television in eight years anyway, so the defect would have no serious consequences. But Corning stood by its policy. They convinced Sony that it was not only Sony's reputation in question, but also Corning's reputation! Sony was very impressed and was very pleased to have such a great supplier.

This example shows how a right policy led to a right work culture. Many companies try to introduce the culture of safety with a very loose connection to policy. No wonder there are so many recalls despite good intent. The next step for management is to periodically make sure the policy works. Then senior management is free to empower employees! Then there is no fear of punishment and employees can speak up without guilt. Management also needs to periodically depend on quality assurance feedback from customers through personal interviews with users such as nurses, doctors, and patients. Written customer complaints are insufficient. Hospitals don't report about 80% of problems. The nurses and doctors are too busy to go to computers and document problem reports.

Management Tasks for Preventing Recalls

Regardless of the form, the policy should set the stage for developing the structure for implementing good design practices and good manufacturing practices. Setting the structure should be the responsibility of the vice president of engineering with the help of the risk manager or the manager in charge of regulatory affairs. If possible, use a cross-functional team that includes all major stakeholders, especially the research and development (R&D) managers, production managers, and marketing managers.

Carolyn Clancy, the former head of the Agency for Healthcare Research and Quality (AHRQ), in her testimony before the

Subcommittee on Health Care, Committee on Finance, U.S. Senate [1] describes a structure for hospitals. A similar thought process can be applied to a medical device company. According to Dr. Clancy,

> A structure represents the basic characteristics of physicians, hospitals, other professionals, and other facilities. It describes whether there are well-educated health professionals, appropriate hospitals, nursing homes, and clinics, as well as well-maintained medical records and good mechanisms for communication between clinicians. For example: Is the mammography equipment up to date and maintained properly? Are the cardiologists well trained and board certified? If the structure is solid, we can concern ourselves with the process of medical care. Concern for process suggests that quality is determined not just by having the right people and facilities available, but also by having the right things get done in the right way.

The structure should be based on similar sound principles of quality. Dr. Clancy defines healthcare quality as getting the right care to the right patient at the right time—every time. The implementation of this vision is a sound structure. In addition, there must be safeguards if an activity is not performed correctly. This definition of quality is very appropriate for medical devices, and can be summarized as follows:

> Quality for a medical device is its ability to provide right care for every user at any demanded time— every time. There must be a safeguard if a device fails to perform its function right.

There is deep meaning in this statement. An example of a patient wearing a pacemaker and what happened to him illustrates the meaning. The patient went to a large department store in a shopping mall [2]. The cashier station of the store had a magnetic device to detect hidden bar codes in expensive items to alert security in case of a theft. The magnetic field in the cashier's station changed the software setting in the pacemaker even though this person stole nothing. He died in front of the cashier because of an excessively high breathing rate. Therefore, according to this definition of quality, the device did not provide right care and there was no safeguard when the device failed to perform its functions right! If the device was designed right, the software engineers should have designed a safe default level in the software for any unusual condition. It is management's responsibility to make sure that policy supports this definition of quality. This definition is not in conflict with regulatory definitions. It strengthens the current thinking on definitions of quality. It includes design for lifetime reliability; that is what the words "every time" imply.

Product Management Procedures

Device safety and high return on investment are not opposite goals if you compare the total cost of doing the right things versus not doing the right things over a period of at least five years. Sometimes it is hard to put numbers on intangible benefits such as getting more customers, avoiding negligence claims, and avoiding patient harm. But a good manager can see them intuitively. Usually, employees already have a cheap and simple solution. Toyota Motors calls such solutions "elegant." The problem lies in lack of good procedures and failure to let teams go in any direction they want. Some written procedures enforcing good engineering management principles should always be in place. This can be the responsibility of product engineering management. The effectiveness of procedures must be verified periodically and a report on how well the procedures are working should be sent to senior management. Examples of such procedures are as follows:

- Product definition procedure
- Risk analysis procedure
- Product development procedure (with flowchart)
- Design review procedure
- Design control procedure
- Requirement analysis procedure
- Software development procedure
- Written performance specifications
- Written interface specifications (for software-dependent products and production processes)
- Integration of regulatory requirements
- Engineering change control procedure
- Design for safety procedure
- Design for reliability procedure
- Design for maintainability procedure
- System integration procedure
- Design verification procedure
- Design validation procedure
- CAPA (corrective action and preventive action) procedure
- Production design verification procedure
- Production validation procedure
- Project control procedure
- Design transfer (to manufacturing) procedure
- Postproduction validation procedure
- Production monitoring procedure

A matrix that shows the owner of each procedure is a plus in managing the integrity of procedures. It is a good idea to brief senior management on the details of each procedure.

Management Reviews

Senior management should be aware of effectiveness reports and make sure manpower and financial resources are adequate. They must thoroughly review risk analysis and risk acceptance decisions. Engineers do tend to make quick decisions in the interest of schedules and costs. Good engineers, with a little brainstorming in a team, almost always find ways to reduce risk and get a high return on investment simultaneously. Senior management should be willing to invest in such elegant solutions. The following example is worth repeating. An anesthesiology device had hoses to deliver oxygen and other gases to patients during surgeries. Occasionally, the surgeons put patients into a coma. They delivered wrong gases to patients because maintenance technicians connected the oxygen hose to a wrong gas outlet. The risk was completely eliminated by designing a unique noninterchangeable fitting for each hose. The product cost went up a little, but the savings were enormous. Customers were very pleased, sales increased for the device maker, and there were no more lawsuits. Zero probability of product recall!

So, what can management do to avoid the pitfalls of making uninformed quick decisions and cutting corners because of short-sightedness? The best strategy management can use is to review the summaries of risk analysis meetings. Find out if the engineers are challenging the design sufficiently. Or, are they voting blindly in the interest of groupthink, as shown in the Figure 13.1?

If they are challenging the design, are they using the ISO 14971 tools such as Preliminary Hazard Analysis (PHA), Failure Mode and Effects Analysis (FMEA), Fault Tree Analysis (FTA), Hazard and Operability (HAZOP) analysis, and Hazard Analysis and Critical Control Points (HACCP) analysis? Senior management members do not need to be experts on these tools, but they should have sufficient understanding.

Monitoring Risk Management Processes

There must be an independent eye for monitoring the risk analysis and risk management processes. Management can

"I said all those in favor say 'Aye.'"

Figure 13.1   Hazards of the groupthink process.

monitor risks themselves by reviewing the risk analysis reports, or by having an independent team assess the effectiveness, or by hiring an outside consultant. To monitor the risk analysis process, management needs to be aware of the following principles:

■ The probability of recall is the product of two probabilities: the probability of the hazard in the device and the probability of the hazardous situation occurring (see ISO 14971 for examples).
■ If the engineering team can prevent either of these probabilities, the recall is no longer a possibility. Harm cannot happen until both the situations occur at the same time.
■ If the above probabilities cannot be designed out, then there should a safeguard to protect the patient. Example: if a patient turns the knob of an infusion device in the wrong direction, the device should be designed to detect the unreasonable demand and limit the dose delivered.

Using Good Paradigms for Efficiency

Efficiency should be a major concern. Why? Because too many engineers waste a tremendous amount of time on trivia and overlook high risks. This author saw a 400-page document on FMEA done by a regulatory affairs engineer after weekly meetings for 6 months with quality assurance staff. The author asked, "What did you do with it?" The answer was shocking. He said, "Nothing." The author asked, "Why wasn't anything done?" The answers were getting worse: "My job is to do the analysis. It is up to design engineers to make design changes." The design engineers had not taken any action.

Management must question what is wrong with this kind of meaningless FMEA. It did not have a procedure on responsibilities and accountability for doing it right and taking timely actions. The second thing that went wrong was that the FMEA was done by a quality assurance engineer who did not fully understand the design. The FMEA should be done by a well-represented cross-functional team that includes design and manufacturing engineers, and the regulatory affairs engineer. The third thing that went wrong was that design improvements were not identified in the document. The fourth thing that went wrong was that the 6-month period for doing the FMEA was too long. Most FMEAs can be done in less than two days by focusing on it for the entire day. When we do the analysis piecemeal, i.e., once a week, we lose focus on important things. Another thing that went wrong was that the team spent too much time on very low risks and not enough time on high risks. Very low risks can be delegated to day-to-day quality assurance teams instead of wasting the time of the entire team. That is why there were 400 pages with no value. An obvious lesson to be learned is that high risks should be handled with urgency with enough time to avoid the risks.

Here is another example of gross inefficiency. The articular surface replacement, or ASR, is used in artificial hips. Some patients, many of whom suffered severe pain and injury from metallic debris generated by the device, spent years trying to convince doctors that there was a problem while the company denied that a problem existed [2]. Traditional artificial hips, which are made of metal and plastic, typically last 15 years or more before requiring replacement. Internal company documents showed that the device would fail within 5 years in 40% of patients. The device should have been recalled immediately when the internal memo was written.

Some paradigms of efficiency are as follows:

■ *Focus on innovation to eliminate risks with simple solutions* [3].

■ *Discourage the groupthink process.* This is a phenomenon that occurs within a group of people, in which the desire for harmony or conformity with the dominant players results in an incorrect or deviant decision-making outcome. Group members try to minimize conflict and reach a consensus decision without critical evaluation of alternative ideas or viewpoints.

■ *Manage teams with the big picture in mind.* Criteria for selecting team members and the consideration given to team composition should start with a clear team goal that defines the boundary of analysis, statement of work, description of the project, responsibilities of each team member, requirements specification, responsibilities for communications with all stakeholders, frequency of reports, and the structure of reports to senior management.

■ *Understand the system.* If everyone does his or her job without fully understanding interdependencies among the complex system interactions, the product is bound to fail. The system is not only about internal people but also interactions with hospital equipment, patients, doctors, nurses, software in the system, and foreseeable misuse. These all have to be integrated with a system mission that is "nothing bad will happen to users."

Summary

Management is responsible for 85% of problems. These problems are created by an inadequate policy, inadequate procedures, and inadequate follow-up. The policy and procedures must be corrected.

References

1. Clancy, Carolyn, MD. "What Is Health Care Quality and Who Decides?" Testimony, the Subcommittee on Health Care, Committee on Finance, U.S. Senate, March 18, 2009, http://www.hhs.gov/asl/testify/2009/03/t20090318b.html
2. Meier, Barry. November 19, 2013, "Johnson & Johnson in Deal to Settle Hip Implant Lawsuits," New York Times, November 19, 2013, http://www.nytimes.com/2013/11/20/business/johnson-johnson-to-offer-2-5-billion-hip-device-settlement.html?_r=1&
3. Raheja, Dev. *Safer Hospital Care: Strategies for Continuous Innovation,* Boca Raton, FL: Taylor & Francis, 2011.

Chapter 14

Innovation Methods Useful in Preventing Recalls

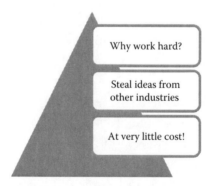

Why work hard?

Steal ideas from other industries

At very little cost!

Introduction

Medical devices are primarily innovative devices. Competition on device performance and innovation is the primary business practice. This is true for innovations in safety. Customers want absolute safety at the lowest price. The goal of innovation is to just do that: high performance and high return on investment.

If we can have high return on investment, we can certainly lower the price. There are several ways we can innovate: stop using outdated ways, use wisdom from experience (heuristics), use classic innovative thinking methods, or use the profound knowledge of the gurus who started the quality revolution.

Stop Using Outdated Practices

If you are not looking for a better idea, you are not going see one, says Malcolm Gladwell, author of *What the Dog Saw: And Other Adventures* (2009). Most organizations get saddled with making their existing products or services incrementally better versus making smart innovations. They become wedded to their existing processes under the guise of best practices and become complacent. One of the best processes used by the medical device industry is Six Sigma for quality and safety improvements. This tool was introduced 30 years ago by Motorola. It was a very good tool at that time, when safety was not taken very seriously. This tool is not appropriate for safety-related defects where the goal has to be zero defects. This tool, first of all, takes about 10 years to implement companywide. It is fine to use this tool for quality improvement, but it does not have sufficient value for preventing recalls.

If it takes 10 years to implement, companies do not want to ship unsafe devices for that many years. Even if a company achieves the Six Sigma level for safety-related defects (target: 3.4 defects per million uses), there can be a very high number of incidents involving major harm to patients, and fatalities. With millions of devices in use each day, which translates to millions of uses each day, there can be too many fatalities each day. If airlines depended only on Six Sigma, there would be about 11 crashes per week at the Chicago airport alone! If hospitals achieve Six Sigma levels, the following scenarios can still exist [1]:

- 76 newborn babies in the United States will be *given to wrong parents each month.*
- At least 1,000 patients will receive the wrong medication per week.
- There would be 107 incorrect medical procedures performed every day.
- There would be no electricity for 10 minutes each week.

Today, safety is the national goal for medical devices. The Joint Commission on Accreditation of Healthcare Organizations, an accreditation agency for hospitals, made medical device alarm safety its number one goal for 2014. The U.S. Food and Drug Administration (FDA) is just as concerned and has started questioning alarm safety in premarket approvals.

The problem is, therefore, finding new-generation tools to help meet alarm safety goals. Actually, there are nice old-generation tools that are still outstanding, but they are not used adequately by many medical device companies. These tools are Preliminary Hazard Analysis (PHA), Fault Tree Analysis (FTA), and Failure Mode and Effects Analysis (FMEA). As mentioned several times in this text, they are recommended in ISO 14971, but most companies are using only FMEA, and many are using it improperly, according to this author's observations. These tools were described in Chapter 2. They are supposed to be proactive, which implies that they should be used for reviewing the entire design and the entire process early in the design. Each time there is a design change or a process change, an FTA should be done on the change.

Another old-generation tool, which is new to medical devices, is the Swiss cheese model of human error. This model is used in hospitals and describes four levels of human failure, each of which influences the next (Figure 14.1). It is very useful in understanding why device defects happen [2], and is an excellent tool for investigating mishaps.

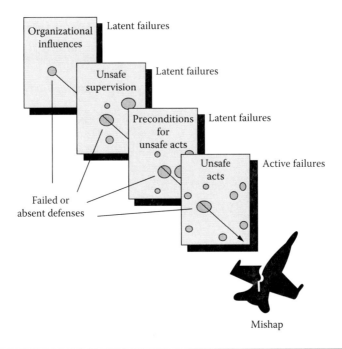

Figure 14.1 **The Swiss cheese model of human error causation. (From Shappell, Scott A., and Wiegmann, Douglas A.** *The Human Factors Analysis and Classification System–HFACS,* **U.S. Department of Transportation, Office of Aviation Medicine, February 2000, http:// www.nifc.gov/safety/reports/humanfactors_class&anly.pdf)**

What makes the Swiss cheese model particularly useful is that it forces investigators to address latent system failures within the causal sequence of events. As their name suggests, latent failures, unlike their active counterparts, may lie dormant or undetected for hours, days, weeks, or even longer, until one day they adversely affect the unsuspecting user. Consequently, they may be overlooked by a management with even the best intentions.

Each management layer that contains hazards is called a *slice of cheese.* Each hazard is called a *hole in the cheese* (hence the name Swiss cheese). The important thing to understand is that harm to the user cannot happen unless the hazards in each layer are simultaneously present. In other words, harm can be prevented if these holes are plugged in a hazard mitigation process. This is good information for innovation in the management process.

Use Heuristics

Preventing adverse events requires special techniques and innovation. Many innovations automatically happen when the mind is trained to think outside the box. Heuristics are a way to push our thinking over the edge. They are short statements of lessons learned, or "rules of thumb," if you will. They may be words of wisdom from experience, educated guesses, or simply innovation. They are often shortcuts in complex situations and can lead to smarter and more efficient solutions. They work most of the time. Over the span of 30 years, this author has learned many lessons:

- Heuristic 1: *If you stop working on unsafe methods, then you are likely to discover safer methods.* Relying on inspection and testing is an unsafe method. Relying on preventing defects by improving the device design is innovation.
- Heuristic 2: *Learn to say "no" to yes-men.* If all team members say "yes" to a new device design or a new process, your answer should be "no." If no one is challenging the design or questioning whether the knowledge, assumptions, and risk are understood well, the design is sure to malfunction or fail. We discussed this paradigm earlier. Challenge the design. The more the challenge, the more the innovation!
- Heuristic 3: *If you want to solve tough problems, steal ideas from other industries.* Critical mishaps must be prevented quickly and efficiently. Over 100 hospitals are following the Toyota Production System where employees are empowered to form a team and solve problems

immediately. If they cannot prevent the problem, a vice president or the head of medicine is contacted through a dedicated telephone line. He or she personally helps the team.

■ Heuristic 4: *If you want quick innovation, create a sense of urgency.* Innovative companies, such as Apple and Google, use this method.

■ Heuristic 5: *No action is an action.* Sometimes, engineers take no action due to lack of time or resources. It may be done out of denial or because of fear of the inability to take action. Whatever the case, the device is usually the victim of the recall. *No action* should not be tolerated.

■ Heuristic 6: *No control is the best control.* We already discussed this. When a device and the process are designed to be defect-free, no control is required. This is the best way to get high quality at low cost.

■ Heuristic 7: *Incident reporting should include reporting on peers, but must be nonpunitive.* Many employees make mistakes in manufacturing and see problems with device design, process design, and serviceability. They should be asked to fill out *Incidence Report* forms and rewarded for reporting their own mistakes and for reporting mistakes by peers (names not necessary). These reports are very productive in preventing hidden risks.

■ Heuristic 8: *Convert bad news into good news.* Device failures and complaints are bad news. Instead of ignoring them, quickly initiate a design change for each. Each design change becomes good news for preventing recalls.

■ Heuristic 9: *Don't test if you already know the product is going to fail.* Engineers typically know in advance that the device will fail durability and reliability tests. They simply want to know how many will fail and how. The same information is already there in the FMEAs, so why waste time and money? Fix the failure modes before conducting tests. Being more efficient adds value to testing.

■ Heuristic 10: *The instant an organization becomes satisfied with its current level of functioning is the moment the organization's quality begins to decline.* Never be satisfied with the current level of quality. There is always a better way, according to Thomas Edison. Every success is a step to a higher success.

Use the Profound Knowledge of the Quality and Safety Gurus

Dr. Edward W. Deming, who was the chief architect for Japan becoming a world leader in quality, used his Theory of Profound Knowledge. According to him, as already discussed, 85% of quality problems are system problems and management is responsible for them. Profound knowledge is made up of four parts: systems thinking, process variation, theory of knowledge, and the psychology of human behavior.

Dr. Deming showed time and again how thinking in terms of a system is critically important for discovering, analyzing, and solving a wide range of problems that businesses continually face. Deming taught that it was the duty of management to inquire and find the answers to performance-related questions. Once the answers are in hand, it is management's duty to take the correct action to improve the system. To do this, Deming taught that there are always variations in systems and there are two specific types of variations: common cause and special cause variations. Common cause variations are built into the system and cause rework, while special cause variations are unique events from outside the system with factors of which management may not be aware. Understanding the difference between the two types of variation and the causes is how management removes problems in the system [3].

The third point, theory of knowledge, challenges us to test opinions, theories, hunches, and beliefs to understand what we know and what we don't. The fourth is about how management must understand that people are different and their motivation can come in many different forms. Deming rejected management-by-carrot-stick rewards, quotas, and merit ratings, which seek to affix blame and to reward individuals. These discourage innovation. Deming urged promoting an environment of trust, interdependence, and pride of workmanship.

Philip Crosby became very famous after he published his book *Quality Is Free* (1979). Crosby's only performance standard is *zero defects*. This is a very appropriate model for medical devices. This is achieved by trying to do the work right the first time. This is what ISO 14971 is about—using the correct tools to predict problems and mitigate them before they happen.

The third guru who transformed the knowledge of safety is James Reason (the mastermind behind the Swiss cheese model), a cognitive psychologist in the United Kingdom. According to Reason, harm to the user is not just about people, it is about the design of the system, which includes equipment and procedures. The solution is to trap errors before they reach the point of harm. Reason explains that accidents happen from a combination of active errors (such as designer errors) and latent errors (such as inadequate design analysis procedures) that are dormant in the system. The more safeguards there are on latent errors, the less likely that mistakes will have a large impact. He uses the analogy of Swiss cheese explained earlier. It is worth going through it again. Assume there are system protections for latent hazards such as good company policy, management involvement, design standards, and knowledge of risk prevention. Each protection system is like a slice of Swiss cheese. A hole is equivalent to a latent hazard sitting there until it is activated due to failure of the protection. There can be several other holes in the protection. Each hole is a path to an adverse event. For a designer or production operator error to cause an adverse event, a hole in each of the above protections has to be activated. In other words, a hole in each protection layer has to let in the disruptive path before harm happens. Plug all the holes in the layers of the system so that the cheese has no holes!

Use Classic Innovation Methods

We can use classic methods for innovation, too. Here are a few [4]:

- Marginal
- Incremental
- Radical
- Disruptive
- Accidental
- Strategic
- Diffusion
- Translocation

The most obvious method is to use incremental innovation. It is an extension of the current design but often with extraordinary results. At the Helen Hamlyn Centre for Design in London [5], a diverse group—consisting of designers, clinicians, psychologists, and human factors experts—studied medical mistakes with an eye toward producing devices that could reduce them. They called the project DOME, short for *designing out medical error.*

The lead designer of DOME was Jonathan West. He quickly realized, while shadowing doctors and nurses at the hospitals of the Imperial College of London, that what makes the problem so persistent is that it's extremely complex. During patient observations, the DOME team noticed practitioners often had to hunt for hand sanitizer, gloves, and aprons. Many had to carry around bins to dispose of needles. Medication cabinets were often blocked or located far from the bedside. And there wasn't always an easy place to scribble notes into a patient chart. Each of these problems is fraught with the potential for a preventable medical error. One of the prototypes they developed ultimately became a device called the CareCentre, which

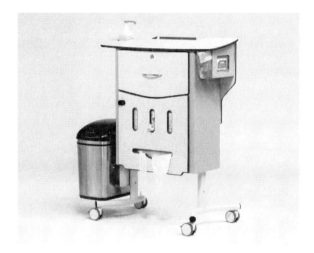

Figure 14.2   The CareCentre innovation. (Courtesy of the Helen Hamlyn Centre for Design.)

serves all these needs in a single, free-standing station that can be situated at the foot of any patient's bed—complete with hand gel, gloves and aprons, drug locker, waste and needle bins, chart surface, and storage slot (Figure 14.2). The team put the device through a week-long randomized trial on surgical wards. Compared to wards with standard equipment, those with the CareCentre had better hand hygiene, fresh glove and apron use, and old glove and apron disposal.

Of special interest to medical device designers would be radical innovations. These are sometimes called *breakthrough innovations*. They result in the creation of a totally new technology or a process that displaces the current way. An example is the da Vinci surgical robot, which has revolutionized the way surgeries are performed. It costs more than $3 million, but its increasing use suggests that the benefits to hospitals and patients are much higher. These innovations create new ways to satisfy customers. Such technologies in the future may replace the need for surgeons in many surgeries. There are other companies introducing robots for knee surgery, hip surgery, and liver surgeries.

The cheapest category is translocation innovation. Steve Jobs, the founder of Apple, used to say, "Good innovators are not good at creating new ideas. They are good at stealing ideas." Apple used this type of innovation. It did not develop the mouse for computers. They stole it from Xerox (they did get permission later from Xerox). Here is another radical innovation from Xerox: a medical device that measures all the vital signs of patients (blood pressure, heart rate, oxygen saturation, respiration rate) without touching the patient with anything!

Summary

Why work hard? Adopt ideas from other industries. There is very little cost and a very high return on investment. There is no need to invent every time. Hospitals have saved thousands of lives and millions of dollars using the Toyota Production System.

References

1. Maryland Hospital Association. *Six Sigma Green Belt Training Manual*, 2009.
2. Shappell, Scott A., and Wiegmann, Douglas A. *The Human Factors Analysis and Classification: System–HFACS*, U.S. Department of Transportation, Office of Aviation Medicine, February 2000, http://www.nifc.gov/safety/reports/ humanfactors_class&anly.pdf
3. Deming Institute, "The System of Profound Knowledge," https://www.deming.org/theman/theories/profoundknowledge
4. Christensen, C. *The Innovator's Dilemma: The Revolutionary Book That Will Change the Way You Do Business*, reprint edition. New York: HarperBusiness, 2011.
5. Jaffe, Eric. "Using Good Design to Eliminate Medical Errors," *Fast Company*, November 2013, http://www.fascodesign. com/3021303/evidence/using-good-design-to-eliminate-medical-errors

Chapter 15

Proactive Role of Marketing in Preventing Recalls

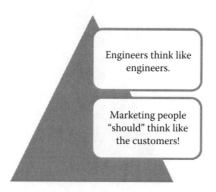

Engineers think like engineers.

Marketing people "should" think like the customers!

Introduction

There are two types of companies: market driven and market driving. A company that is *market driven* is controlled by the expressed needs of the customers. A company that is *marketing driving* is one that anticipates future needs of customers and aims to consistently market safety and reliability features. A marketing-driving company says, "If you like the device, tell your friends. If you don't like it, tell me." Companies need to be marketing driving to not only reduce the chance of recall but also to use safety as a marketing tool for high return on investment.

Don't Repeat Failures of Yesterday in Devices of Tomorrow

"It's not enough to do your best. You must first know what to do, and then do your best." This is the most important lesson Dr. W. Edwards Deming taught. This teaching still has not hit home. Most companies have several repeat failure modes and customer complaints that were made with respect to previous products. Marketing can make a big difference by taking the past histories of design reviews and making suggestions for device improvement.

Why are companies prone to having past problems in new devices? One of the main reasons seems to be *weak cognition*. It is usually the result of short-term memory, inattention to detail, insufficient vigilance, or multitasking. This unsafe work propagates unknowingly because of disengaged employees, insufficient understanding of customer needs, splitting design among several designers, and ineffective risk analysis. Another big reason is that ineffective device management procedures often touted as "best practices" are like a bad virus with a long incubation period and no early warning, often resulting in sudden catastrophic malfunction of the device. Most companies still hold individuals responsible for mistakes instead of fixing the management process.

Marketing can help to prevent the recurrence of past mistakes and improve efficiency of the device design. They have records of customer complaints for years. They even have knowledge of what problems exist on similar devices because they hear about them from customers during marketing efforts. They can develop a Recall Prevention Checklist (RPC) of complaints and device failures before a new device specification is approved. In addition, the checklist can include the problems and failures of competitive devices. In other words, they can help develop a device that delights customers!

Gather Intelligence on Customer Safety Needs during Lead Generation

Customers don't usually have a good feel for safety problems until the device gives them problems or malfunctions. But they do know the problems that prompted them to buy your device versus your competitor's. When potential customers look at your device, they typically look for more than your sales information. They want to know third-party opinions of your device, your reputation for safety, and how you differ from other manufacturers. This intelligence gathering can be useful in strengthening the device RPC. This list reminds all engineers that the aim of new devices is to prevent recalls proactively instead fixing problems one at a time later.

Gather Intelligence on Safety and Quality Issues on Device Search Engines

An Internet search on similar devices can provide valuable information on customer preferences and gripes. The U.S. Food and Drug Administration (FDA) has good websites for complaints and adverse events, such as the following:

- *Manufacturer and User Facility Device Experience (MAUDE):* http://www.accessdata.fda.gov/scripts/cdrh/cfdocs/cfmaude/textsearch.cfm
- *MedSun Reports on medical device failure and malfunction reports:* http://www.accessdata.fda.gov/scripts/cdrh/cfdocs/Medsun/searchReportText.cfm

- *Postapproval studies:* http://www.fda.gov/medicaldevices/
  deviceregulationandguidance/postmarketrequirements/
  postapprovastudies/ucm2005741.htm
- *Postmarket surveillance studies:* http://www.accessdata.
  fda.gov/scripts/cdrh/cfdocs/cfpma/pss.cfm (accessed March
  29, 2014).
- *List of device recalls:* http://www.fda.gov/medicaldevices/
  safety/ListofRecalls/default.htm

Participate in Design Reviews to Be an Advocate for Users

Engineers think like engineers and marketing people *should* think like the customer. If you are designing a product to be used by nurses, then get nurses on your design team. Simple things like naming conventions, color, and where a button is located are big deals—they may not cause a recall but they will impact sales. This author can think of many examples of how products could have been designed better if the customer or people close to the customer got involved earlier. A director of nursing at a premier hospital told this author that she had invited the engineers of a device to improve the design to make it safer for patients. After hearing the complaints, the engineers came up with several excellent ideas, but she was surprised that they had practically no knowledge of the problems the nurses were facing.

Therefore, there is an urgent need for a paradigm shift in medical device safety to enhance quality, safety, and reliability through powerful applications of human factor engineering to optimize the relationship between users and medical devices, and monitor interactions among devices. The marketing department must act as a guide and lead companies to bring awareness of variations in use/misuse and device interactions with other devices. This communication is vital to improving the design process.

Marketing has to participate in design reviews much more than once. Even sales and applications/sales training people should participate in this process. Figure 15.1 shows the entire chain of device design and production reviews [1]. Depending on the risk analysis, marketing should choose where to participate, including review of validations.

The design feasibility stage should be the first stop for marketing. In this stage, solutions are generated to meet design requirements. There should be more than one design concept. That is where marketing contributes most. They can compare their design to competitors' designs from a different perspective than the design engineer. A variety of methods for concept development exist such as the Six Thinking Hats [2] developed by Dr. Edward de Bono. The aim is to generate as many ideas as possible. Ideas should come from marketing also since they know the customers as well as the competition. The range of concept designs can then be systematically chosen to determine the most suitable concept.

The concept design may involve the following:

■ Simple sketches of the concept
■ Functional Failure Mode and Effects Analysis (FMEA) after choosing the idea
■ Risk analysis on the selected idea
■ Review of initial manufacturing concept to predict manufacturing defects

Marketing should definitely participate in the detailed design review to verify that the risks have been eliminated or reduced to an acceptable level. Many techniques, such as FMEA and Fault Tree Analysis (FTA) can be used to assess the efficacy and efficiency of the device. FMEA can have a number of variations that address different aspects of the device. Examples are the design FMEA, the process FMEA, and the user FMEA. Marketing should lead the effort on the user FMEA. The aim is to thoroughly review what can go wrong

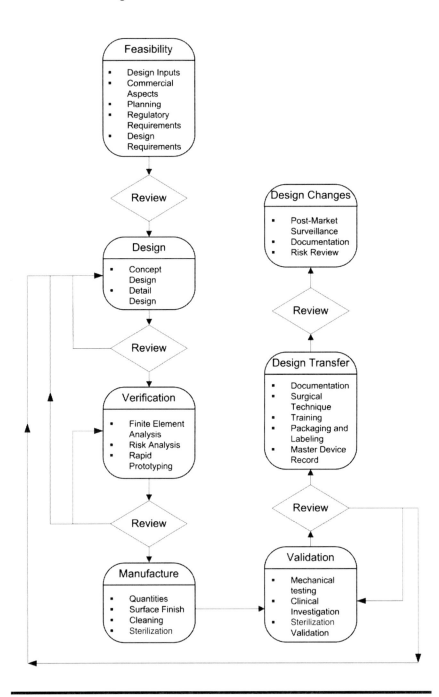

Figure 15.1   Device design and manufacturing reviews. (http://www .bentham-science.com/open/tobej/articles/V003/21TOBEJ.pdf)

in use and what kinds of misuse might occur. The device manufacturer must consider how to protect patients in case of misuse of the device, such as in patient-controlled analgesia (PCA) pumps. Many patients have died from inadvertent misuse of these devices. It is also important not to isolate the medical device itself from the packaging, sterilization, and labeling functions, or the possibility that the wrong kits might be shipped. These areas have many risks associated with them that could adversely affect the performance of the device.

Review New Device Specifications with Trusted Customers

Some companies invite trusted and loyal customers to participate in the concept design review. This is one of the best approaches for preventing recalls, but customers must be briefly trained in risk prediction and risk analysis. Customers should definitely be on the user FMEA team since they have firsthand knowledge of potential uses and misuses.

If it is not possible for customers (especially doctors) to participate in the user FMEA, marketing can visit trusted doctors who have a vested interest in the device and personally review uses and misuses of the device with them.

Provide Intelligence in Risk Assessment to Ensure the Public Health Benefits Outweigh the Risk

When the risks are not acceptable, but the device is needed in the interest of public health (such as computed tomography [CT] scanners), marketing should be the key player in deciding whether to market the device. Everyone knows there are high risks in CT scanners (206 patients at one hospital were over-radiated for more than 6 months in 2009) but the FDA allows marketing of these devices because it saves thousands of lives each year.

Marketing can help because it has the knowledge of who and how many users in the public domain will benefit from the device and how much they will benefit. In addition, marketing has the means to collect such data.

Market Safety Features to Promote the Device and to Get Feedback from Users

Customers usually want to discuss safety concerns with marketing. Marketing should encourage them to do so. The more marketing knows about customer concerns, the less chance there is of a device recall. One way to collect customer concerns is to post testimonials from doctors and nurses on the device company's website and ask caregivers to comment. This is a productive starting point in the discussions during lead generations.

Summary

There is an urgent need for a paradigm shift in medical device safety to enhance quality, safety, and reliability through powerful application of human factor engineering to optimize the relationship between users and medical devices, and interactions among devices. The marketing department must help guide the design process.

References

1. Aitchison, G.A. et al. "A Review of the Design Process for Implantable Orthopedic Medical Devices," *Open Biomedical Engineering Journal* (online), July 2, 2009, U.S. National Library of Medicine, National Institutes of Health, http://www. benthamscience.com/open/tobej/articles/V003/21TOBEJ.pdf
2. de Bono, Edward. *Six Thinking Hats*. London: Penguin, 1990.

Appendix A: Medical Device Safety from the Hospital's Point of View

Protecting Patients from Hidden Dangers in Medical Devices

Dev Raheja, MS, CSP, Patient System Safety, and
Peter P. Pronovost, MD, PhD, Johns Hopkins University

[Reprinted from *Focus on Patient Safety* 13, no. 2 (2010), the quarterly newsletter of the National Patient Safety Foundation]

Dangers to patients from medical devices lurk in large numbers. These dangers are present in many devices, such as: MRI equipment, CT scanners, nuclear medicine equipment, defibrillators, infusion pumps, ventilators patient monitoring systems, electronic health record computers. There are also dangers in medical accessories such as: interface connectors, surgical trays, syringes and valves, IV tubes that transport blood, catheters. While medical devices are obviously designed to benefit patients, they can also cause harm. Healthcare professionals at all levels must be made aware of these high risks. In 2008, the U.S. Food and Drug Administration (FDA) recalled 845 medical devices—the highest number ever recalled in a single year [1].

From 1999 to 2005, one in three FDA device recalls were due to software errors [2]. This article addresses some examples of hidden dangers in medical devices and offers several suggestions for how hospitals can work to reduce these dangers.

What Are the Dangers?

The hidden dangers in medical devices and accessories come from a number of sources:

Inherent technology. Radiation from CT scans performed in 2007 was estimated to cause 29,000 cases of cancer and kill nearly 15,000 Americans, according to findings published in the *Archives of Internal Medicine* and reported in the *New York Times* [3]. Cardiologist Rita Redberg, an *Archives* editor, states, "What we learned is there is a significant amount of radiation with these CT scans, more than what we thought [4]."

Equipment software. In 2005, a Florida hospital disclosed that 77 brain cancer patients had received 50% more radiation than prescribed because one of the most powerful—and supposedly precise—linear accelerators had been programmed incorrectly for nearly a year [5]. Software usage failures have resulted in delays or interruptions in the operation of infusion pumps, interfering with intervention and causing serious harm [6].

User interface. The computerized physician order entry (CPOE) interface has resulted in prescribing errors, improper dosage or quantity, wrong dosage form, extra doses, omission errors, unauthorized or wrong drugs, and drugs administered to the wrong patient [7]. In one emergency department's color-coded screen, a colorblind nurse had problems interpreting priorities.

Device reliability. Medical devices are designed and assembled by humans. Because of variability in human factors as well as in the manufacturing process, the devices can fail at any time during normal use. In some portable ventricular assist devices (VADs), the drivers stopped when the compressor motor wore out much sooner than after the expected

3,000 hours of use. A compressor motor can stop without warning. When the motor fails, there is a loss of VAD support for the patient, causing inadequate blood flow to and from the heart [8].

Defective components. Some pacemakers were recalled when defective leads caused serious injuries. Pinholes and exposed wire braids were found in some catheters, which could result in a brain clot or a blood vessel puncture. The catheters were recalled.

Sneak conditions. Sometimes devices deliver unexpected behavior due to so-called sneak circuits—especially electronic devices, which are subject to uncontrolled interactions among circuits, software, and components. The devices can fail to deliver required functions, deliver wrong functions, deliver unexpected outputs, fail to deliver a function when they should, or deliver too much or too little treatment. There are many instances of such failures in ventilators, defibrillators, and infusion pumps. A hospital reported a fatal central venous air embolism caused by the separation of a specific manufacturer's side port/hemostasis valve catheter-to-sheath adapter from the same manufacturer's percutaneous sheath hub. The accident occurred when the standard Luer-lock fitting disconnected as a patient was moved from a bed to a chair. The hospital could have bought a one-piece device interface instead of a 2-piece interface from the same manufacturer to avoid the mishap [9].

Expired sterility. Infections can result from surgical devices with an expired sterility date, or those packaged in unsterilized or improperly sanitized bags.

Accessories. The FDA has warned that the plastic material in IV tubes used in cardiac surgery may contain BPA, a highly toxic chemical. BPA can combine with warm blood in the tubes.

Incorrect labeling information or instructions. The label insert for a video-guided catheter contained inappropriate information on use with an energy-delivering instrumentation.

Instructions for an implantable ventricular assist device (IVAD) stated it could be implanted or placed in the external position. If the IVAD is placed externally, air leaks might develop in the pneumatic driveline [10].

Incorrect servicing. Accidents from inaccurate settings of MRI machines or improper calibrations of other devices are not uncommon. FDA recall Z-0165-2008, *Class 2 Recall ONCOR Expression,* suggests that the miscalibration can affect the beam profile of the radiation therapy system [11].

These are only some of the dangers inherent in medical devices. For a listing of medical device recalls, visit the FDA's web site, www.fda.gov.

How Can Hospitals Protect Patients from These Dangers?

Once a device has been approved by the FDA, current law protects medical device companies from failure-to-warn and design-defect lawsuits on defects designed into that device. To help protect patients from potential dangers in medical devices and accessories, hospitals can: use FDA recall information to strengthen the current device use procedures and develop safeguards and checklists; conduct a hazard analysis such as the Failure Mode and Effects Analysis for dangers on new technologies and devices. Some hazards will not be on the FDA recall list, as the dangers have not shown up yet; discuss the results of the hazard analysis with suppliers so they can design the equipment to include safeguards, alerts, monitoring software for inaccurate performance, and built-in self-checks. An example of a built-in check is a car's software that makes sure the air bag is in good working condition every time the ignition key is turned. Hospitals can create the healthcare equivalent of the Commercial Aviation Safety Team (CAST), a public–private partnership to reduce hazards in aviation. Peter Pronovost and colleagues are exploring a healthcare version of CAST with an ad hoc group whose stakeholders include the Agency for Healthcare Research and Quality (AHRQ), the FDA, the Joint Commission, ECRI Institute, and more than 15 large health systems. This approach has been named the Public Private Partnership to Promote Patient Safety (P5S) [12]. Hospitals can use knowledge of inherent medical device and accessory dangers in bedside intelligence systems.

Use a Team Approach to Risk Reduction

Patient safety safeguards and barriers should be in place in every area, including medical device use, where a possibility of harm exists. The care givers may have to exercise due diligence. Considering that many patients are exposed to nearly all the medical device dangers discussed above, the probabilities can add up, resulting in highly significant risks. Risk managers and patient safety officers should work together to reduce these risks.

References

1. Problems related to software, heparin help push recalls to all time high. *"The Silver Sheet."* 2009(Feb)13(2):1–12. Available at: *http://offers.fdcreports.com/silver/freeissue.pdf.* Accessed June 10, 2010.
2. Anderson P. Safety critical coding standards reduce medical device risks. *Med Design Technology,* 2009(Feb/Mar); 13(2):68,70. Available at: *http://e-ditionsbyfry.com/activemagazine/welcome/mdt/mdt090301.asp.* Accessed April 8, 2010.
3. Berrington de González A, Mahesh M, Kim KP, et al. Projected cancer risks from computed tomographic scans performed in the United States in 2007. *Arch Intern Med.* 2009(Dec14);169(22):2071–2077.
4. 15,000 will die from CT scans done in 1 year. Reuters. Dec 14, 2009. Available at: *http://www.msnbc.msn.com/id/34420356/ns/health-cancer/.* Accessed April 8, 2010.

5. Bogdanich W. Radiation offers new cures, and ways to do harm. *New York Times.* January 24, 2010:A1. Available at: http://www.nytimes.com/2010/01/24/health/24radiation.html, Accessed April 8, 2010.

6. *Baxter Colleague Single and Triple Channel Volumetric Infusion Pumps – Class 1 Recall.* Silver Spring, MD: U.S. Food and Drug Administration; January 23, 2009. Available at: *http://www.fda.gov/MedicalDevices/Safety/ RecallsCorrectionsRemovals/ListofRecalls/ucm117583.htm.* Accessed April 8, 2010.*npsf.org.*

7. Santell J. *Computer-related Errors: What Every Pharmacist Should Know* [presentation]. American Society of Health-System Pharmacists Midyear Clinical Meeting. Orlando, FL, December 9, 2004. Available at: *http://www.usp.org/pdf/EN/ patientSafety/slideShows2004-12-09.pdf.* Accessed April 8, 2010.

8. *Thoratec® TLC-ll® Portable Ventricular Assist Device (VAD) Driver – Class 1 Recall.* Silver Spring, MD: U.S. Food and Drug Administration; June 8, 2007. Available at: *http://www.fda.gov/ MedicalDevices/Safety/RecallsCorrectionsRemovals/ListofRecalls/ ucm062465.htm.* Accessed April 8, 2010.

9. Medical Device Safety Reports database. Plymouth Meeting, PA: The ECRI Institute. Available at: *http://www.mdsr.ecri.org/ Default.aspx.* Accessed April 8, 2010.

10. *Thoratec Corporation Implantable Ventricular Assist Devices (IVAD) – Class 1 Recall.* Silver Spring, MD: U.S. Food and Drug Administration; October 19, 2007. Available at: *http://www. fda.gov/MedicalDevices/Safety/RecallsCorrectionsRemovals/ ListofRecalls/ucm062374.htm.* Accessed April 8, 2010.

11. *ONCOR Expression Medical Charge-particle Radiation Therapy System – Class 2 Recall.* Silver Spring, MD: U.S. Food and Drug Administration; November 9, 2007. Available at: http://www. *accessdata.fda.gov/scripts/cdrh/cfdocs/cfres/res.cfm?id = 65672.* Accessed July 9, 2010.

12. Pronovost PJ, Goeschel CA, Olsen KL, et al. Reducing health care hazards: lessons from the Commercial Aviation Safety Team. *Health Affairs (Millwood).* 2009, Available at: *http:// dx.doi.org/10.1377/hlthaff.28.3.w479.* Accessed April 8, 2008.

Appendix B: The FDA Quality System Regulation

Code of Federal Regulations Title 21 (Food and Drugs), Part 820

Subpart A—General Provisions

Sec. 820.1 Scope.

(a) *Applicability.* (1) Current good manufacturing practice (CGMP) requirements are set forth in this quality system regulation. The requirements in this part govern the methods used in, and the facilities and controls used for, the design, manufacture, packaging, labeling, storage, installation, and servicing of all finished devices intended for human use. The requirements in this part are intended to ensure that finished devices will be safe and effective and otherwise in compliance with the Federal Food, Drug, and Cosmetic Act (the act). This part establishes basic requirements applicable to manufacturers of finished medical devices. If a manufacturer engages in only some operations subject to the requirements in this part, and not in others, that manufacturer need only comply with those requirements applicable to the operations in which it is engaged. With respect to class I devices, design controls apply only to those devices listed in 820.30(a)(2).

This regulation does not apply to manufacturers of components or parts of finished devices, but such manufacturers are encouraged to use appropriate provisions of this regulation as guidance. Manufacturers of human blood and blood components are not subject to this part, but are subject to part 606 of this chapter. Manufacturers of human cells, tissues, and cellular and tissue-based products (HCT/Ps), as defined in 1271.3(d) of this chapter, that are medical devices (subject to premarket review or notification, or exempt from notification, under an application submitted under the device provisions of the act or under a biological product license application under section 351 of the Public Health Service Act) are subject to this part and are also subject to the donor-eligibility procedures set forth in part 1271 subpart C of this chapter and applicable current good tissue practice procedures in part 1271 subpart D of this chapter. In the event of a conflict between applicable regulations in part 1271 and in other parts of this chapter, the regulation specifically applicable to the device in question shall supersede the more general.

(2) The provisions of this part shall be applicable to any finished device as defined in this part, intended for human use, that is manufactured, imported, or offered for import in any State or Territory of the United States, the District of Columbia, or the Commonwealth of Puerto Rico.

(3) In this regulation the term "where appropriate" is used several times. When a requirement is qualified by "where appropriate," it is deemed to be "appropriate" unless the manufacturer can document justification otherwise. A requirement is "appropriate" if nonimplementation could reasonably be expected to result in the product not meeting its specified requirements or the manufacturer not being able to carry out any necessary corrective action.

(b) The quality system regulation in this part supplements regulations in other parts of this chapter except where explicitly stated otherwise. In the event of a conflict between applicable regulations in this part and in other parts of this chapter, the regulations specifically applicable to the device in question shall supersede any other generally applicable requirements.

(c) *Authority.* Part 820 is established and issued under authority of sections 501, 502, 510, 513, 514, 515, 518, 519, 520, 522, 701, 704, 801, 803 of the act (21 U.S.C. 351, 352, 360, 360c, 360d, 360e, 360h, 360i, 360j, 360l, 371, 374, 381, 383). The failure to comply with any applicable provision in this part renders a device adulterated under section 501(h) of the act. Such a device, as well as any person responsible for the failure to comply, is subject to regulatory action.

(d) *Foreign manufacturers.* If a manufacturer who offers devices for import into the United States refuses to permit or allow the completion of a Food and Drug Administration (FDA) inspection of the foreign facility for the purpose of determining compliance with this part, it shall appear for purposes of section 801(a) of the act, that the methods used in, and the facilities and controls used for, the design, manufacture, packaging, labeling, storage, installation, or servicing of any devices produced at such facility that are offered for import into the United States do not conform to the requirements of section 520(f) of the act and this part and that the devices manufactured at that facility are adulterated under section 501(h) of the act.

(e) *Exemptions or variances.* (1) Any person who wishes to petition for an exemption or variance from any device quality system requirement is subject to the requirements of section 520(f)(2) of the act. Petitions for an exemption or variance shall be submitted according to the procedures set forth in 10.30 of this chapter, the FDA's administrative procedures. Guidance is available from the Food and

Drug Administration, Center for Devices and Radiological Health, Division of Small Manufacturers, International and Consumer Assistance, 10903 New Hampshire Ave., Bldg. 66, rm. 4613, Silver Spring, MD 20993-0002, 1-800-638-2041 or 301-796-7100, FAX: 301-847-8149.

(2) FDA may initiate and grant a variance from any device quality system requirement when the agency determines that such variance is in the best interest of the public health. Such variance will remain in effect only so long as there remains a public health need for the device and the device would not likely be made sufficiently available without the variance.

[61 FR 52654, Oct. 7, 1996, as amended at 65 FR 17136, Mar. 31, 2000; 65 FR 66636, Nov. 7, 2000; 69 FR 29829, May 25, 2005; 72 FR 17399, Apr. 9, 2007; 75 FR 20915, Apr. 22, 2010]

Sec. 820.3 Definitions.

(a) *Act* means the Federal Food, Drug, and Cosmetic Act, as amended (secs. 201-903, 52 Stat. 1040 et seq., as amended (21 U.S.C. 321-394)). All definitions in section 201 of the act shall apply to the regulations in this part.

(b) *Complaint* means any written, electronic, or oral communication that alleges deficiencies related to the identity, quality, durability, reliability, safety, effectiveness, or performance of a device after it is released for distribution.

(c) *Component* means any raw material, substance, piece, part, software, firmware, labeling, or assembly which is intended to be included as part of the finished, packaged, and labeled device.

(d) *Control number* means any distinctive symbols, such as a distinctive combination of letters or numbers, or both, from which the history of the manufacturing, packaging, labeling, and distribution of a unit, lot, or batch of finished devices can be determined.

(e) *Design history file (DHF)* means a compilation of records which describes the design history of a finished device.

(f) *Design input* means the physical and performance requirements of a device that are used as a basis for device design.

(g) *Design output* means the results of a design effort at each design phase and at the end of the total design effort. The finished design output is the basis for the device master record. The total finished design output consists of the device, its packaging and labeling, and the device master record.

(h) *Design review* means a documented, comprehensive, systematic examination of a design to evaluate the adequacy of the design requirements, to evaluate the capability of the design to meet these requirements, and to identify problems.

(i) *Device history record (DHR)* means a compilation of records containing the production history of a finished device.

(j) *Device master record (DMR)* means a compilation of records containing the procedures and specifications for a finished device.

(k) *Establish* means define, document (in writing or electronically), and implement.

(l) *Finished device* means any device or accessory to any device that is suitable for use or capable of functioning, whether or not it is packaged, labeled, or sterilized.

(m) *Lot or batch* means one or more components or finished devices that consist of a single type, model, class, size, composition, or software version that are manufactured under essentially the same conditions and that are intended to have uniform characteristics and quality within specified limits.

(n) *Management with executive responsibility* means those senior employees of a manufacturer who have the authority to establish or make changes to the manufacturer's quality policy and quality system.

(o) *Manufacturer* means any person who designs, manufactures, fabricates, assembles, or processes a finished device. Manufacturer includes but is not limited to those who perform the functions of contract sterilization, installation, relabeling, remanufacturing, repacking, or specification development, and initial distributors of foreign entities performing these functions.

(p) *Manufacturing material* means any material or substance used in or used to facilitate the manufacturing process, a concomitant constituent, or a byproduct constituent produced during the manufacturing process, which is present in or on the finished device as a residue or impurity not by design or intent of the manufacturer.

(q) *Nonconformity* means the nonfulfillment of a specified requirement.

(r) *Product* means components, manufacturing materials, in-process devices, finished devices, and returned devices.

(s) *Quality* means the totality of features and characteristics that bear on the ability of a device to satisfy fitness-for-use, including safety and performance.

(t) *Quality audit* means a systematic, independent examination of a manufacturer's quality system that is performed at defined intervals and at sufficient frequency to determine whether both quality system activities and the results of such activities comply with quality system procedures, that these procedures are implemented effectively, and that these procedures are suitable to achieve quality system objectives.

(u) *Quality policy* means the overall intentions and direction of an organization with respect to quality, as established by management with executive responsibility.

(v) *Quality system* means the organizational structure, responsibilities, procedures, processes, and resources for implementing quality management.

(w) *Remanufacturer* means any person who processes, conditions, renovates, repackages, restores, or does any other
act to a finished device that significantly changes the
finished device's performance or safety specifications, or
intended use.

(x) *Rework* means action taken on a nonconforming product so that it will fulfill the specified DMR requirements
before it is released for distribution.

(y) *Specification* means any requirement with which a product, process, service, or other activity must conform.

(z) *Validation* means confirmation by examination and provision of objective evidence that the particular requirements
for a specific intended use can be consistently fulfilled.
(1) *Process validation* means establishing by objective evidence that a process consistently produces a result or
product meeting its predetermined specifications.
(2) Design validation means establishing by objective
evidence that device specifications conform with user
needs and intended use(s).

(aa) *Verification* means confirmation by examination and provision of objective evidence that specified requirements
have been fulfilled.

Sec. 820.5 Quality System.

Each manufacturer shall establish and maintain a quality
system that is appropriate for the specific medical device(s)
designed or manufactured, and that meets the requirements of
this part.

Subpart B—Quality System Requirements

Sec. 820.20 Management Responsibility.

(a) *Quality policy.* Management with executive responsibility shall establish its policy and objectives for, and

commitment to, quality. Management with executive responsibility shall ensure that the quality policy is understood, implemented, and maintained at all levels of the organization.

(b) *Organization.* Each manufacturer shall establish and maintain an adequate organizational structure to ensure that devices are designed and produced in accordance with the requirements of this part.

   (1) *Responsibility and authority.* Each manufacturer shall establish the appropriate responsibility, authority, and interrelation of all personnel who manage, perform, and assess work affecting quality, and provide the independence and authority necessary to perform these tasks.

   (2) Resources. Each manufacturer shall provide adequate resources, including the assignment of trained personnel, for management, performance of work, and assessment activities, including internal quality audits, to meet the requirements of this part.

   (3) Management representative. Management with executive responsibility shall appoint, and document such appointment of, a member of management who, irrespective of other responsibilities, shall have established authority over and responsibility for:

     (i) Ensuring that quality system requirements are effectively established and effectively maintained in accordance with this part; and

     (ii) Reporting on the performance of the quality system to management with executive responsibility for review.

(c) *Management review.* Management with executive responsibility shall review the suitability and effectiveness of the quality system at defined intervals and with sufficient frequency according to established procedures to ensure that the quality system satisfies the requirements of this part and the manufacturer's established quality policy and objectives. The dates and results of quality system reviews shall be documented.

(d) *Quality planning.* Each manufacturer shall establish a quality plan which defines the quality practices, resources, and activities relevant to devices that are designed and manufactured. The manufacturer shall establish how the requirements for quality will be met.

(e) *Quality system procedures.* Each manufacturer shall establish quality system procedures and instructions. An outline of the structure of the documentation used in the quality system shall be established where appropriate.

Sec. 820.22 Quality Audit.

Each manufacturer shall establish procedures for quality audits and conduct such audits to assure that the quality system is in compliance with the established quality system requirements and to determine the effectiveness of the quality system. Quality audits shall be conducted by individuals who do not have direct responsibility for the matters being audited. Corrective action(s), including a reaudit of deficient matters, shall be taken when necessary. A report of the results of each quality audit, and reaudit(s) where taken, shall be made and such reports shall be reviewed by management having responsibility for the matters audited. The dates and results of quality audits and reaudits shall be documented.

Sec. 820.25 Personnel.

(a) *General.* Each manufacturer shall have sufficient personnel with the necessary education, background, training, and experience to assure that all activities required by this part are correctly performed.

(b) *Training.* Each manufacturer shall establish procedures for identifying training needs and ensure that all personnel are trained to adequately perform their assigned responsibilities. Training shall be documented.

(1) As part of their training, personnel shall be made aware of device defects which may occur from the improper performance of their specific jobs.

(2) Personnel who perform verification and validation activities shall be made aware of defects and errors that may be encountered as part of their job functions.

Subpart C—Design Controls

Sec. 820.30 Design Controls.

(a) *General.* (1) Each manufacturer of any class III or class II device, and the class I devices listed in paragraph (a)(2) of this section, shall establish and maintain procedures to control the design of the device in order to ensure that specified design requirements are met.

(2) The following class I devices are subject to design controls:

(i) Devices automated with computer software; and

(ii) The devices listed in the following chart.

Section	Device
868.6810	Catheter, Tracheobronchial Suction.
878.4460	Glove, Surgeon's.
880.6760	Restraint, Protective.
892.5650	System, Applicator, Radionuclide, Manual.
892.5740	Source, Radionuclide Teletherapy.

(b) *Design and development planning.* Each manufacturer shall establish and maintain plans that describe or reference the design and development activities and define responsibility for implementation. The plans shall identify and describe the interfaces with different groups or activities that provide, or result in, input to the design and development process. The plans shall be reviewed, updated, and approved as design and development evolves.

(c) *Design input.* Each manufacturer shall establish and maintain procedures to ensure that the design requirements relating to a device are appropriate and address the intended use of the device, including the needs of the user and patient. The procedures shall include a mechanism for addressing incomplete, ambiguous, or conflicting requirements. The design input requirements shall be documented and shall be reviewed and approved by a designated individual(s). The approval, including the date and signature of the individual(s) approving the requirements, shall be documented.

(d) *Design output.* Each manufacturer shall establish and maintain procedures for defining and documenting design output in terms that allow an adequate evaluation of conformance to design input requirements. Design output procedures shall contain or make reference to acceptance criteria and shall ensure that those design outputs that are essential for the proper functioning of the device are identified. Design output shall be documented, reviewed, and approved before release. The approval, including the date and signature of the individual(s) approving the output, shall be documented.

(e) *Design review.* Each manufacturer shall establish and maintain procedures to ensure that formal documented reviews of the design results are planned and conducted at appropriate stages of the device's design development. The procedures shall ensure that participants at each design review include representatives of all functions concerned with the design stage being reviewed and an individual(s) who does not have direct responsibility for the design stage being reviewed, as well as any specialists needed. The results of a design review, including identification of the design, the date, and the individual(s) performing the review, shall be documented in the design history file (the DHF).

(f) *Design verification.* Each manufacturer shall establish and maintain procedures for verifying the device design. Design verification shall confirm that the design output meets the design input requirements. The results of the design verification, including identification of the design, method(s), the date, and the individual(s) performing the verification, shall be documented in the DHF.

(g) *Design validation.* Each manufacturer shall establish and maintain procedures for validating the device design. Design validation shall be performed under defined operating conditions on initial production units, lots, or batches, or their equivalents. Design validation shall ensure that devices conform to defined user needs and intended uses and shall include testing of production units under actual or simulated use conditions. Design validation shall include software validation and risk analysis, where appropriate. The results of the design validation, including identification of the design, method(s), the date, and the individual(s) performing the validation, shall be documented in the DHF.

(h) *Design transfer.* Each manufacturer shall establish and maintain procedures to ensure that the device design is correctly translated into production specifications.

(i) *Design changes.* Each manufacturer shall establish and maintain procedures for the identification, documentation, validation or where appropriate verification, review, and approval of design changes before their implementation.

(j) *Design history file.* Each manufacturer shall establish and maintain a DHF for each type of device. The DHF shall contain or reference the records necessary to demonstrate that the design was developed in accordance with the approved design plan and the requirements of this part.

Subpart D—Document Controls

Sec. 820.40 Document Controls.

Each manufacturer shall establish and maintain procedures to control all documents that are required by this part. The procedures shall provide for the following:

(a) *Document approval and distribution.* Each manufacturer shall designate an individual(s) to review for adequacy and approve prior to issuance all documents established to meet the requirements of this part. The approval, including the date and signature of the individual(s) approving the document, shall be documented. Documents established to meet the requirements of this part shall be available at all locations for which they are designated, used, or otherwise necessary, and all obsolete documents shall be promptly removed from all points of use or otherwise prevented from unintended use.

(b) *Document changes.* Changes to documents shall be reviewed and approved by an individual(s) in the same function or organization that performed the original review and approval, unless specifically designated otherwise. Approved changes shall be communicated to the appropriate personnel in a timely manner. Each manufacturer shall maintain records of changes to documents. Change records shall include a description of the change, identification of the affected documents, the signature of the approving individual(s), the approval date, and when the change becomes effective.

Subpart E—Purchasing Controls

Sec. 820.50 Purchasing Controls.

Each manufacturer shall establish and maintain procedures to ensure that all purchased or otherwise received product and services conform to specified requirements.

(a) *Evaluation of suppliers, contractors, and consultants.* Each manufacturer shall establish and maintain the requirements, including quality requirements, that must be met by suppliers, contractors, and consultants. Each manufacturer shall:

   (1) Evaluate and select potential suppliers, contractors, and consultants on the basis of their ability to meet specified requirements, including quality requirements. The evaluation shall be documented.

   (2) Define the type and extent of control to be exercised over the product, services, suppliers, contractors, and consultants, based on the evaluation results.

   (3) Establish and maintain records of acceptable suppliers, contractors, and consultants.

(b) *Purchasing data.* Each manufacturer shall establish and maintain data that clearly describe or reference the specified requirements, including quality requirements, for purchased or otherwise received product and services. Purchasing documents shall include, where possible, an agreement that the suppliers, contractors, and consultants agree to notify the manufacturer of changes in the product or service so that manufacturers may determine whether the changes may affect the quality of a finished device. Purchasing data shall be approved in accordance with 820.40.

Subpart F—Identification and Traceability

Sec. 820.60 Identification.

Each manufacturer shall establish and maintain procedures for identifying product during all stages of receipt, production, distribution, and installation to prevent mixups.

Sec. 820.65 Traceability.

Each manufacturer of a device that is intended for surgical implant into the body or to support or sustain life and whose failure to perform when properly used in accordance with instructions for use provided in the labeling can be reasonably expected to result in a significant injury to the user shall establish and maintain procedures for identifying with a control number each unit, lot, or batch of finished devices and where appropriate components. The procedures shall facilitate corrective action. Such identification shall be documented in the DHR.

Subpart G—Production and Process Controls

Sec. 820.70 Production and Process Controls.

(a) *General.* Each manufacturer shall develop, conduct, control, and monitor production processes to ensure that a device conforms to its specifications. Where deviations from device specifications could occur as a result of the manufacturing process, the manufacturer shall establish and maintain process control procedures that describe any process controls necessary to ensure conformance to specifications. Where process controls are needed they shall include:

(1) Documented instructions, standard operating procedures (SOP's), and methods that define and control the manner of production;

(2) Monitoring and control of process parameters and component and device characteristics during production;

(3) Compliance with specified reference standards or codes;

(4) The approval of processes and process equipment; and

(5) Criteria for workmanship which shall be expressed in documented standards or by means of identified and approved representative samples.

(b) *Production and process changes.* Each manufacturer shall establish and maintain procedures for changes to a specification, method, process, or procedure. Such changes shall be verified or where appropriate validated according to 820.75, before implementation and these activities shall be documented. Changes shall be approved in accordance with 820.40.

(c) *Environmental control.* Where environmental conditions could reasonably be expected to have an adverse effect on product quality, the manufacturer shall establish and maintain procedures to adequately control these environmental conditions. Environmental control system(s) shall be periodically inspected to verify that the system, including necessary equipment, is adequate and functioning properly. These activities shall be documented and reviewed.

(d) *Personnel.* Each manufacturer shall establish and maintain requirements for the health, cleanliness, personal practices, and clothing of personnel if contact between such personnel and product or environment could reasonably be expected to have an adverse effect on product quality. The manufacturer shall ensure that maintenance and other personnel who are required to work temporarily under special environmental conditions are appropriately trained or supervised by a trained individual.

(e) *Contamination control.* Each manufacturer shall establish and maintain procedures to prevent contamination of equipment or product by substances that could reasonably be expected to have an adverse effect on product quality.

(f) *Buildings.* Buildings shall be of suitable design and contain sufficient space to perform necessary operations, prevent mixups, and assure orderly handling.

(g) *Equipment.* Each manufacturer shall ensure that all equipment used in the manufacturing process meets specified requirements and is appropriately designed, constructed, placed, and installed to facilitate maintenance, adjustment, cleaning, and use.

  (1) *Maintenance schedule.* Each manufacturer shall establish and maintain schedules for the adjustment, cleaning,

and other maintenance of equipment to ensure that manufacturing specifications are met. Maintenance activities, including the date and individual(s) performing the maintenance activities, shall be documented.

(2) Inspection. Each manufacturer shall conduct periodic inspections in accordance with established procedures to ensure adherence to applicable equipment maintenance schedules. The inspections, including the date and individual(s) conducting the inspections, shall be documented.

(3) Adjustment. Each manufacturer shall ensure that any inherent limitations or allowable tolerances are visibly posted on or near equipment requiring periodic adjustments or are readily available to personnel performing these adjustments.

(h) *Manufacturing material.* Where a manufacturing material could reasonably be expected to have an adverse effect on product quality, the manufacturer shall establish and maintain procedures for the use and removal of such manufacturing material to ensure that it is removed or limited to an amount that does not adversely affect the device's quality. The removal or reduction of such manufacturing material shall be documented.

(i) *Automated processes.* When computers or automated data processing systems are used as part of production or the quality system, the manufacturer shall validate computer software for its intended use according to an established protocol. All software changes shall be validated before approval and issuance. These validation activities and results shall be documented.

Sec. 820.72 Inspection, Measuring, and Test Equipment.

(a) *Control of inspection, measuring, and test equipment.* Each manufacturer shall ensure that all inspection, measuring, and test equipment, including mechanical, automated, or electronic inspection and test equipment, is suitable for its intended purposes and is capable of producing valid results. Each manufacturer shall establish and maintain procedures to ensure that equipment is routinely calibrated, inspected, checked, and maintained. The procedures shall include provisions for handling, preservation, and storage of equipment, so that its accuracy and fitness for use are maintained. These activities shall be documented.

(b) *Calibration.* Calibration procedures shall include specific directions and limits for accuracy and precision. When accuracy and precision limits are not met, there shall be provisions for remedial action to reestablish the limits and to evaluate whether there was any adverse effect on the device's quality. These activities shall be documented.

　(1) *Calibration standards.* Calibration standards used for inspection, measuring, and test equipment shall be traceable to national or international standards. If national or international standards are not practical or available, the manufacturer shall use an independent reproducible standard. If no applicable standard exists, the manufacturer shall establish and maintain an in-house standard.

　(2) Calibration records. The equipment identification, calibration dates, the individual performing each calibration, and the next calibration date shall be documented. These records shall be displayed on or near each piece of equipment or shall be readily available to the personnel using such equipment and to the individuals responsible for calibrating the equipment.

Sec. 820.75 Process Validation.

(a) Where the results of a process cannot be fully verified by subsequent inspection and test, the process shall be validated with a high degree of assurance and approved according to established procedures. The validation activities and results, including the date and signature of the individual(s) approving the validation and where appropriate the major equipment validated, shall be documented.

(b) Each manufacturer shall establish and maintain procedures for monitoring and control of process parameters for validated processes to ensure that the specified requirements continue to be met.

   (1) Each manufacturer shall ensure that validated processes are performed by qualified individual(s).

   (2) For validated processes, the monitoring and control methods and data, the date performed, and, where appropriate, the individual(s) performing the process or the major equipment used shall be documented.

(c) When changes or process deviations occur, the manufacturer shall review and evaluate the process and perform revalidation where appropriate. These activities shall be documented.

Subpart H—Acceptance Activities

Sec. 820.80 Receiving, In-Process, and Finished Device Acceptance.

(a) *General.* Each manufacturer shall establish and maintain procedures for acceptance activities. Acceptance activities include inspections, tests, or other verification activities.

(b) *Receiving acceptance activities.* Each manufacturer shall establish and maintain procedures for acceptance of incoming product. Incoming product shall be inspected, tested, or otherwise verified as conforming to specified requirements. Acceptance or rejection shall be documented.

(c) *In-process acceptance activities.* Each manufacturer shall establish and maintain acceptance procedures, where appropriate, to ensure that specified requirements for in-process product are met. Such procedures shall ensure that in-process product is controlled until the required inspection and tests or other verification activities have been completed, or necessary approvals are received, and are documented.

(d) *Final acceptance activities.* Each manufacturer shall establish and maintain procedures for finished device acceptance to ensure that each production run, lot, or batch of finished devices meets acceptance criteria. Finished devices shall be held in quarantine or otherwise adequately controlled until released. Finished devices shall not be released for distribution until:

(1) The activities required in the DMR are completed;

(2) the associated data and documentation is reviewed;

(3) the release is authorized by the signature of a designated individual(s); and

(4) the authorization is dated.

(e) *Acceptance records.* Each manufacturer shall document acceptance activities required by this part. These records shall include:

(1) The acceptance activities performed;

(2) the dates acceptance activities are performed;

(3) the results;

(4) the signature of the individual(s) conducting the acceptance activities; and

(5) where appropriate the equipment used. These records shall be part of the DHR.

Sec. 820.86 Acceptance Status.

Each manufacturer shall identify by suitable means the acceptance status of product, to indicate the conformance or nonconformance of product with acceptance criteria. The identification of acceptance status shall be maintained throughout manufacturing, packaging, labeling, installation, and servicing of the product to ensure that only product which has passed the required acceptance activities is distributed, used, or installed.

Subpart I—Nonconforming Product

Sec. 820.90 Nonconforming Product.

(a) *Control of nonconforming product.* Each manufacturer shall establish and maintain procedures to control product that does not conform to specified requirements. The procedures shall address the identification, documentation, evaluation, segregation, and disposition of nonconforming product. The evaluation of nonconformance shall include a determination of the need for an investigation and notification of the persons or organizations responsible for the nonconformance. The evaluation and any investigation shall be documented.

(b) *Nonconformity review and disposition.* (1) Each manufacturer shall establish and maintain procedures that define the responsibility for review and the authority for the disposition of nonconforming product. The procedures shall set forth the review and disposition process. Disposition of nonconforming product shall be documented. Documentation shall include the justification for use of nonconforming product and the signature of the individual(s) authorizing the use.

(2) Each manufacturer shall establish and maintain procedures for rework, to include retesting and reevaluation of the nonconforming product after rework, to ensure that the product meets its current approved specifications. Rework and reevaluation activities, including a determination of any adverse effect from the rework upon the product, shall be documented in the DHR.

Subpart J—Corrective and Preventive Action

Sec. 820.100 Corrective and Preventive Action.

(a) Each manufacturer shall establish and maintain procedures for implementing corrective and preventive action.
The procedures shall include requirements for:

(1) Analyzing processes, work operations, concessions,
quality audit reports, quality records, service records,
complaints, returned product, and other sources of
quality data to identify existing and potential causes
of nonconforming product, or other quality problems.
Appropriate statistical methodology shall be employed
where necessary to detect recurring quality problems;

(2) Investigating the cause of nonconformities relating to
product, processes, and the quality system;

(3) Identifying the action(s) needed to correct and prevent
recurrence of nonconforming product and other quality problems;

(4) Verifying or validating the corrective and preventive
action to ensure that such action is effective and does
not adversely affect the finished device;

(5) Implementing and recording changes in methods and
procedures needed to correct and prevent identified
quality problems;

(6) Ensuring that information related to quality problems
or nonconforming product is disseminated to those
directly responsible for assuring the quality of such
product or the prevention of such problems; and

(7) Submitting relevant information on identified quality
problems, as well as corrective and preventive actions,
for management review.

(b) All activities required under this section, and their results,
shall be documented.

Subpart K—Labeling and Packaging Control

Sec. 820.120 Device Labeling.

Each manufacturer shall establish and maintain procedures to control labeling activities.

(a) *Label integrity.* Labels shall be printed and applied so as to remain legible and affixed during the customary conditions of processing, storage, handling, distribution, and where appropriate use.

(b) *Labeling inspection.* Labeling shall not be released for storage or use until a designated individual(s) has examined the labeling for accuracy including, where applicable, the correct expiration date, control number, storage instructions, handling instructions, and any additional processing instructions. The release, including the date and signature of the individual(s) performing the examination, shall be documented in the DHR.

(c) *Labeling storage.* Each manufacturer shall store labeling in a manner that provides proper identification and is designed to prevent mixups.

(d) *Labeling operations.* Each manufacturer shall control labeling and packaging operations to prevent labeling mixups. The label and labeling used for each production unit, lot, or batch shall be documented in the DHR.

(e) *Control number.* Where a control number is required by 820.65, that control number shall be on or shall accompany the device through distribution.

Sec. 820.130 Device Packaging.

Each manufacturer shall ensure that device packaging and shipping containers are designed and constructed to protect the device from alteration or damage during the customary conditions of processing, storage, handling, and distribution.

Subpart L—Handling, Storage, Distribution, and Installation

Sec. 820.140 Handling.

Each manufacturer shall establish and maintain procedures to ensure that mixups, damage, deterioration, contamination, or other adverse effects to product do not occur during handling.

Sec. 820.150 Storage.

(a) Each manufacturer shall establish and maintain procedures for the control of storage areas and stock rooms for product to prevent mixups, damage, deterioration, contamination, or other adverse effects pending use or distribution and to ensure that no obsolete, rejected, or deteriorated product is used or distributed. When the quality of product deteriorates over time, it shall be stored in a manner to facilitate proper stock rotation, and its condition shall be assessed as appropriate.

(b) Each manufacturer shall establish and maintain procedures that describe the methods for authorizing receipt from and dispatch to storage areas and stock rooms.

Sec. 820.160 Distribution.

(a) Each manufacturer shall establish and maintain procedures for control and distribution of finished devices to ensure that only those devices approved for release are distributed and that purchase orders are reviewed to ensure that ambiguities and errors are resolved before devices are released for distribution. Where a device's fitness for use or quality deteriorates over time, the procedures shall ensure that expired devices or devices deteriorated beyond acceptable fitness for use are not distributed.

(b) Each manufacturer shall maintain distribution records which include or refer to the location of:
(1) The name and address of the initial consignee;
(2) The identification and quantity of devices shipped;
(3) The date shipped; and
(4) Any control number(s) used.

Sec. 820.170 Installation.

(a) Each manufacturer of a device requiring installation shall establish and maintain adequate installation and inspection instructions, and where appropriate test procedures. Instructions and procedures shall include directions for ensuring proper installation so that the device will perform as intended after installation. The manufacturer shall distribute the instructions and procedures with the device or otherwise make them available to the person(s) installing the device.

(b) The person installing the device shall ensure that the installation, inspection, and any required testing are performed in accordance with the manufacturer's instructions and procedures and shall document the inspection and any test results to demonstrate proper installation.

Subpart M—Records

Sec. 820.180 General Requirements.

All records required by this part shall be maintained at the manufacturing establishment or other location that is reasonably accessible to responsible officials of the manufacturer and to employees of FDA designated to perform inspections. Such records, including those not stored at the inspected establishment, shall be made readily available for review and copying by FDA employee(s). Such records shall be legible and shall be stored to minimize deterioration and to prevent loss. Those records stored in automated data processing systems shall be backed up.

(a) *Confidentiality.* Records deemed confidential by the manufacturer may be marked to aid FDA in determining whether information may be disclosed under the public information regulation in part 20 of this chapter.

(b) *Record retention period.* All records required by this part shall be retained for a period of time equivalent to the design and expected life of the device, but in no case less than 2 years from the date of release for commercial distribution by the manufacturer.

(c) *Exceptions.* This section does not apply to the reports required by 820.20(c) Management review, 820.22 Quality audits, and supplier audit reports used to meet the requirements of 820.50(a) Evaluation of suppliers, contractors, and consultants, but does apply to procedures established under these provisions. Upon request of a designated employee of FDA, an employee in management with executive responsibility shall certify in writing that the management reviews and quality audits required under this part, and supplier audits where applicable, have been performed and documented, the dates on which they were performed, and that any required corrective action has been undertaken.

Sec. 820.181 Device Master Record.

Each manufacturer shall maintain device master records (DMR's). Each manufacturer shall ensure that each DMR is prepared and approved in accordance with 820.40. The DMR for each type of device shall include, or refer to the location of, the following information:

(a) Device specifications including appropriate drawings, composition, formulation, component specifications, and software specifications;

(b) Production process specifications including the appropriate equipment specifications, production methods, production procedures, and production environment specifications;

(c) Quality assurance procedures and specifications including acceptance criteria and the quality assurance equipment to be used;

(d) Packaging and labeling specifications, including methods and processes used; and

(e) Installation, maintenance, and servicing procedures and methods.

Sec. 820.184 Device History Record.

Each manufacturer shall maintain device history records (DHR's). Each manufacturer shall establish and maintain procedures to ensure that DHR's for each batch, lot, or unit are maintained to demonstrate that the device is manufactured in accordance with the DMR and the requirements of this part. The DHR shall include, or refer to the location of, the following information:

(a) The dates of manufacture;

(b) The quantity manufactured;

(c) The quantity released for distribution;

(d) The acceptance records which demonstrate the device is manufactured in accordance with the DMR;

(e) The primary identification label and labeling used for each production unit; and

(f) Any device identification(s) and control number(s) used.

Sec. 820.186 Quality System Record.

Each manufacturer shall maintain a quality system record (QSR). The QSR shall include, or refer to the location of, procedures and the documentation of activities required by this part that are not specific to a particular type of device(s), including, but not limited to, the records required by 820.20. Each manufacturer shall ensure that the QSR is prepared and approved in accordance with 820.40.

Sec. 820.198 Complaint Files.

(a) Each manufacturer shall maintain complaint files. Each manufacturer shall establish and maintain procedures for receiving, reviewing, and evaluating complaints by a formally designated unit. Such procedures shall ensure that:
  (1) All complaints are processed in a uniform and timely manner;
  (2) Oral complaints are documented upon receipt; and
  (3) Complaints are evaluated to determine whether the complaint represents an event which is required to be reported to FDA under part 803 of this chapter, Medical Device Reporting.

(b) Each manufacturer shall review and evaluate all complaints to determine whether an investigation is necessary. When no investigation is made, the manufacturer shall maintain a record that includes the reason no investigation was made and the name of the individual responsible for the decision not to investigate.

(c) Any complaint involving the possible failure of a device, labeling, or packaging to meet any of its specifications shall be reviewed, evaluated, and investigated, unless such investigation has already been performed for a similar complaint and another investigation is not necessary.

(d) Any complaint that represents an event which must be reported to FDA under part 803 of this chapter shall be promptly reviewed, evaluated, and investigated by a designated individual(s) and shall be maintained in a separate portion of the complaint files or otherwise clearly identified. In addition to the information required by 820.198(e), records of investigation under this paragraph shall include a determination of:

(1) Whether the device failed to meet specifications;

(2) Whether the device was being used for treatment or diagnosis; and

(3) The relationship, if any, of the device to the reported incident or adverse event.

(e) When an investigation is made under this section, a record of the investigation shall be maintained by the formally designated unit identified in paragraph (a) of this section. The record of investigation shall include:

(1) The name of the device;

(2) The date the complaint was received;

(3) Any device identification(s) and control number(s) used;

(4) The name, address, and phone number of the complainant;

(5) The nature and details of the complaint;

(6) The dates and results of the investigation;

(7) Any corrective action taken; and

(8) Any reply to the complainant.

(f) When the manufacturer's formally designated complaint unit is located at a site separate from the manufacturing establishment, the investigated complaint(s) and the record(s) of investigation shall be reasonably accessible to the manufacturing establishment.

(g) If a manufacturer's formally designated complaint unit is located outside of the United States, records required by this section shall be reasonably accessible in the United States at either:

(1) A location in the United States where the manufacturer's records are regularly kept; or

(2) The location of the initial distributor.

[61 FR 52654, Oct. 7, 1996, as amended at 69 FR 11313, Mar. 10, 2004; 71 FR 16228, Mar. 31, 2006]

Subpart N—Servicing

Sec. 820.200 Servicing.

(a) Where servicing is a specified requirement, each manufacturer shall establish and maintain instructions and procedures for performing and verifying that the servicing meets the specified requirements.

(b) Each manufacturer shall analyze service reports with appropriate statistical methodology in accordance with 820.100.

(c) Each manufacturer who receives a service report that represents an event which must be reported to FDA under part 803 of this chapter shall automatically consider the report a complaint and shall process it in accordance with the requirements of 820.198.

(d) Service reports shall be documented and shall include:
   (1) The name of the device serviced;
   (2) Any device identification(s) and control number(s) used;
   (3) The date of service;
   (4) The individual(s) servicing the device;
   (5) The service performed; and
   (6) The test and inspection data.

[61 FR 52654, Oct. 7, 1996, as amended at 69 FR 11313, Mar. 10, 2004]

Subpart O—Statistical Techniques

Sec. 820.250 Statistical Techniques.

(a) Where appropriate, each manufacturer shall establish and maintain procedures for identifying valid statistical techniques required for establishing, controlling, and verifying the acceptability of process capability and product characteristics.

(b) Sampling plans, when used, shall be written and based on a valid statistical rationale. Each manufacturer shall establish and maintain procedures to ensure that sampling methods are adequate for their intended use and to ensure that when changes occur the sampling plans are reviewed. These activities shall be documented.

Last Updated: 04/01/2013, C:\Documents and Settings\draheja\My Documents\BookMedDevAppendixA.mht

Index